"FlowCycle is the best of JIT, Kaizen, and even Lean Manufacturing. Why? When I met these folks we were just looking for help with a Pull system. We got much more than a Kanban/Pull system. We got a super-charged culture reducing cycle time in every area of our business. What was the main difference? All of FMA's people have real-life experiences in management, engineering, and operations. They have 'walked the talk', unlike other firms we interviewed who wanted to give us a report. FMA was out there with our people, making it happen."

William J. Dignan
Vice President of Operations
Computing Devices International

"I was very impressed with Flow Management's ability to get everyone involved. Even the employees who are normally wall flowers were attentive and even active in our discussions."

Union Steward
Tenneco Packaging

"FlowCycle is the cornerstone of our Lean strategy to produce the highest quality product and increase plant capacity. John Ballis provided the creative educational process to transfer knowledge which led to our 70% reduction in cycle time."

Bob Wagoner
Marine Engines

"We have not only paid for this effort, but we are well on the way to achieving a 10:1 payback in just our fifth month of working with FlowCycle Methodologies."

Ed Opperud
Vice President, Coins Division
De La Rue Cash Systems

"The expertise provided by the FlowCycle Methodologies was invaluable to our Cycle Time Reduction project. The results of the methodologies were extremely positive, allowing us to proceed as rapidly as possible with full implementation. As a result of this methodology we were able to process more custom orders with the same number of people."

Waukesha Bearing Team
Results: 65% Cycle Time Reduction
in Journal & Thrust Bearings

"Wondering if John Ballis's FlowCycle Methodology would really work, we took two assembly lines and implemented MARI. The methodology of MARI very quickly proved that it could do exactly what he said it could. In about six weeks, we saw a 10% productivity improvement throughout the factory, and, on those two assembly lines specifically, we improved about 30%."

Bret Olson
President & CEO
Black & Decker, Brazil

"FlowCycle MARI Methodology is the best thing that has happened to our company in the last five years."

Neil Palmquist
Production Manager
General Dynamics
Information Systems

"Our Lean efforts in manufacturing were guided by John Ballis. We were very pleased with the results of our Cycle Time Reductions and process improvement in Quality and Flow."

Don Cipriano
Plant Manager
Re-Union Cabinets, Inc

MANAGING FLOW

Also by John P. Ballis

TQM – III, An Alternative To Reengineering

MANAGING⚡FLOW

Achieving Lean in the New Millennium to Win the Gold

John P. Ballis

BROWN BOOKS
Dallas, Texas

For more information or to order additional books, please write:

Flow Management & Associates, Inc.

12222 Merit Dr., Suite 1890

Dallas, Texas 75251 U.S.A.

972-239-0689   FAX: 972-239-5315

First Printing 2001

ISBN   0-9654620-0-6

Brown Books

www.brownbooks.com

DEDICATION

To my wife, Sally, my rock and tower of inspiration,
and my wonderful children, John and Heather.
Thank you for standing by me during the times
of building up the company and while
writing this book. Your support and prayers have been
my strength and willingness to stay the course.

"FlowCycle has brought to Unicorn Industries a methodology of process improvements that have advanced our business. FlowCycle approach has led us to benefits that were realized in a timely manner and exceeded our goals, as well as enhanced our customer service level. The methodology is based on people who have 'walked the talk – not just talked the talk'."

Henry W. Mower, *CEO*
Lullabye Corporation

"FlowCycle MARI Methodology has been instrumental in awakening the creative skills of our organization. Throughout the introduction of their structured methodology and through coaching our employees in process redesign techniques, we have made significant improvements in human effectiveness and increased awareness of the Total Business Process."

Boise Cascade
Steering Committee Team

"We worked with FlowCycle MARI Methodology for four months to help us with our Employee Involvement and Empowerment efforts. We are very happy with the results. We have never had so many positive responses from our employees. They have been very successful in changing and helping our people."

Ron Zimmerman
Human Relations Manager
Tenneco Packaging - Tomahawk Mill

"The cross functional team at Don Simon Homes and Marshall Towne Millworks reduced material waste and improved field service by utilizing FlowCycle methodology."

Michelle Stellner
Quality Assurance Manager
Don Simon Homes

"This is the first time I have ever had employees making comments on their ISO sheets! They all loved the MARI training and are waiting in line to take part in future workshops."

Jerry Lowe
Mill Manager
Tenneco Packaging

TABLE OF CONTENTS

Preface xiii

Acknowledgements xvii

PART I: MANAGING FLOW TO ACHIEVE LEAN 1

Introduction 2

Chapter One

A Waste-Less Enterprise™ or Lean Manufacturing? 10

FlowCycle™ – Managing Flow Every day 12

What Is a Workflow? 19

How Workflows Fit in the Value Stream 21

Improving Flow. 22

One-Piece Flow 23

FlowCycle – The MARI™ Execution Methodology 26

The FlowCycle Approach to Achieve Continuous

Improvement 28

The Measures of World Class to Maintain Continuous

Improvement 30

Workflow Cycle Efficiency (WCE) 30

Six Key Measures That Drive Improvement 34

Employee Motivation and Employee Involvement 41

The FlowCycle Advantage 44

FlowCycle/Lean Workflow Improvement. 45

FlowCycle Company Can Out-perform the Competition. 46

FlowCycle Builds on Cycles of Learning 50

Chapter Two

The Three Key Principles of FlowCycle to Achieving a Waste-Less Enterprise . 52

Principle One – Total Quality Culture (TQC) 53

Change Skills – Not People 56

Hallmarks of TQC Business Quality Processes (BQP) 58

Principle Two – Business Quality Processes (BQP) 61

Principle Three – Continuous Quality Manufacturing (CQM) 65

PART II: FlowCycle MARI Methodology 69

MARI – Organizing and Implementing Change. 70

Chapter Three

FlowCycle Phase I – Mobilization 88

Step 1: Build Foundation for Successful Process Change 90

Understanding Change 91

Steering Committee Team Level. 94

Management vs. Leadership 96

Functions of the Steering Committee. 97

The Vision Supported Mission Tasks Statements -
Core Values/Strategy 97

Worldwide Vision of Cycle Time Reduction To Achieve Lean 99

Mission Task Statements – Not Sub-Vision Statements 100

Core Values 101

Core Strategies 102

Core Team. 102

Sector Teams. 103

Implementation Teams 105

Team Ground Rules. 106

Accountability Partners and the Four Styles of Behavior 108

Organizing for Success 111

Step 2: Select the Processes for Redesign 112

Order Fullfillment Processes. 114

Step 3: Launch Change Management Efforts 117

The Red Flag Process - For Guaranteeing Results. 120

Mobilization Conclusion 121

The Off-Site Team Building Experience 126

Mobilization Summary 127

Chapter Four

FlowCycle Phase II – Assessment 129

Step 1: Analyze Customer Requirements 133

Who Is Your Customer ... Really? 134

Surveying Customers. 135

External Customers 135

Measured Customer Satisfaction 135

Internal Customers. 136

Step 2: Assess Current Performance 137

Value Stream Map. 137

CQM/BQP Centers 138

Activity Worksheet 144

Step 3: Conduct Benchmarking 148

Step 4: Analyze Business for Leverage 152

Using a Pareto Chart 153

Managing Resistance During Assessment. 154

Assessment Summary 155

Chapter Five

FlowCycle Phase III – Redesign. 156

Step 1: Set Design Goals and Priorities 158

Setting Improvement Goals 159

Step 2: Create Further Redesign 163

      Net Requirements. 165

Step 3: Document Pilot for Change 169

      Performance Plans 172

      Redesign Summary. 176

Chapter Six

FlowCycle Phase IV – Implementation 177

      Step 1: Develop Detailed Implementation Plans. 179

      Step 2: Pilot and Rollout Test. 181

      Step 3: Rollout and Continuously Improve 186

      Implementation Summary 190

Conclusion 191

      FlowCycle To Success 192

      A Final Word 194

Appendix

      View of a FlowCycle Company 195

Glossary 199

Index. 206

PREFACE

As we enter a new millennium, businesses face a rapidly changing New Economy. The world economy can shift from outstanding to weak in a single quarter, according to one Fortune 500 executive. High-growth economic conditions create an atmosphere of rapid expansion, and weak conditions often result in wide-ranging workforce reductions. Properly managing these extremes requires companies to manage the flow of an often forgotten yet vital part of any company: its workflow processes.

Over the last thirty years, global corporations undertook many initiatives to improve their processes and reduce cycle times. These initiatives produced improvements, but the results were inconsistent and generally not comprehensive. The problem is that those initiatives are not comprehensive strategies for sustained process improvements that can lead to profitable long-term growth.

FlowCycle and its MARI improvement process is an "execution methodology" for successful change, not once, but every day. The methodology creates a path to continually remove waste from every process to achieve the goal of long-term profitable growth and expansion through increased speed and competitiveness. How does FlowCycle accomplish this? By involving every employee of an enterprise in the journey for improvement every hour of every day.

Alan Greenspan, Chairman of the Federal Reserve, gave technology a broad endorsement in early 2001 for its role in the recent high growth rates of the United States economy. According to Greenspan, between the early 1970s and 1995, labor productivity grew about 1 1/2 percent per year on average.

Since 1995, productivity growth has doubled due to advances in the application of information technology to the supply chain, which increased speed and managed inventory levels.

However, the speed and intensity of the slowdown in the economy in the last half of 2000 caught many by surprise. Companies' inventory levels began to balloon, requiring them to make severe adjustments to production rates. Greenspan testified, "Because the extent of the slowdown was not anticipated by businesses, it induced some backup in inventories, despite the more advanced, Just-In-Time (JIT) technologies that have in recent years enabled firms to adjust production levels more rapidly to changes in demand." The point is that improvements to date through JIT and software applications were not enough. What more can businesses do?

Greenspan went on to say that the benefits of applying these new technologies had largely been realized and that to maintain above-average growth rates businesses must focus again on the human aspect of productivity. Companies must get back to the basics and increase the efficiency of manufacturing workflow processes by reducing waste in human efforts too. Applying Internet and information technologies are not the whole answer. Managing and continually improving business and manufacturing workflow processes every single day is the answer. How is this achieved? By involving people.

My introduction to workflow management came in 1984 when IBM sent four other industrial engineers and myself to Japan to learn JIT. We developed a process called Continuous Flow Manufacturing that in 1986 took the assembly of the PS-2 motherboard from 45 days to 4.5 days in a nine-month span. In 1994 I founded Flow Management & Associates, Inc., to coach businesses in reducing cycle times and improving workflow processes in all areas of the business, not just in manufacturing. Workflow management is almost exclusively associated with manufacturing, but FlowCycle is applicable to any type of business, service industries as well as manufacturing companies.

FlowCycle, the MARI Execution methodology, applies to improving processes in your personal life as well. Every process has a flow, and that flow is cycled or repeated. FlowCycle teaches people how to take the waste out of any process.

I hope you enjoy this book and experience a new way of thinking about creating change in your business and your life.

ACKNOWLEDGMENTS

Acknowledgments are the most difficult part of a book to write: it is hard to know where to start and even harder to know where to stop. The central thesis of this book is that global corporations must understand a mandate of Cycle Time Reduction for every value stream to achieve "Waste-Less Enterprise." I credit every CEO, president, manager, supervisor, office staff, and especially hourly employees of my acquaintance over the last 25 years. I owe a great deal to these people from whom I gained knowledge to grow my company. The development of this execution methodology called FlowCycle™-MARI™ could not have evolved without the candid feedback of customers and friends. These conversations started me on a six-year journey, implementing Cycle Time Reduction under the banner of MARI. Your faith in my vision of world class through a simple concept of management practices of change process has yielded the amazing organizational growth of many individuals and companies. I am grateful and indebted to many people who helped me make this book possible. First, I would like to thank all the employees of Flow Management & Associates, Inc. and FlowCycle Inc. who parented the book from its early drafts to completion, adding wisdom, encouragement, and first-rate editorial comments. I am indebted to three special people on the drafts: Pete Hall for perseverance, Allen Fischer on the editorial work, and Janet Dahlheim for bringing the illustrations to life. My gratitude also goes to Brown Books for their collaborative publishing style that contributed to the conceptualization of this book, and their leader, Milli Brown, who works outside the box to make dreams come true.

PART ONE

MANAGING FLOW
TO ACHIEVE LEAN

INTRODUCTION ☙ PART ONE

How does a business manage the flow?

How does a business manage the flow in its value streams to maximize return and minimize effort? Exactly how does a long-time business remove years of waste in its processes, allowing them to flow freely? Further, how does a business that has been operating for years remove that waste in a few short months? Finally, how is non-value-added designed out of a new building or plant up front?

That is what this book is about. After reading this short book, leaders of businesses, from Fortune 500 CEOs to Mom and Pop sole proprietors, will have an easily understood and applicable "Execution Methodology" to guide the successful transformation from a business characterized by systemic waste and frustration to a business characterized by straight-line customer responsiveness and energized, fulfilled employees. This book is not about academic theory. This book is about taking organized, systematic, relentless action. An "Execution Methodology" is a guide or a playbook for accomplishing a goal. In this case, the goal is managing workflow processes for optimum output to achieve speed and stay one step ahead of the competition.

- IBM was called Big Blue for a reason. This computer giant in world-class competition was automatically seen as the king to overthrow. In the late eighties, IBM began losing its competitiveness to Compaq Computers. Then a small, Lean-running machine started by Michael Dell was able to leap ahead in the race and outperform Big Blue and Compaq in the personal computer market.

How did this happen?

Clue: It had nothing to do with the quantity of machines Dell could manufacture and deliver.

- Lt. General Gus Pagonis believes catastrophic failures happen during war because government supply and logistic processes can not support short cycle changes. Typical government procedure is to order, then produce what is necessary. During the Gulf War, soldiers were sent to Wal-Mart, KMart and other stores across the nation to buy long johns because no supplier could deliver what was needed fast enough.

Why did this happen?

Clue: It had nothing to do with the shortage of long johns.

- With the simple act of clicking a fountain pen, a man named Mr. Shingo demonstrated that the only value-added time in the process of assembling an ink pen was the actual pushing of the plunger down to apply ink to paper.

What did that demonstration prove?

Clue: It had nothing to do with the design or price of the pen.

If you are unable to answer these questions, your business may become an example in a future edition of this book. What these stories illustrate is the need for change in today's constantly evolving world. You probably wonder how IBM and Dell relate to your company. What does General Pagonis sending troops to Wal-Mart tell you about the need for change? What does Mr. Shingo's statement, "There was no room for waste between being the strongest and the weakest," have to do with your company? All businesses have waste. That waste makes them weak when the market quickly changes.

International business is a track race with no visible finish line. Companies try to be first to market new products, services, and innovations to get or stay ahead of the competition. Your goal is to win the race before your competitors even get out of the starting blocks.

If you succeed in getting there first, you can't rest on your laurels. To stay ahead of the competition, it's on to the next race … and the next race … and the next race. The competition may feel like a sprint or a marathon, depending on your place in the world market and the state of the economy. After every race, you start training all over again, hoping to win the next one.

In actuality, they are all relay races to complete the supply value chain while you try to meet your customers' needs with flawless handoffs. Winning any single leg of the race doesn't guarantee victory – it doesn't get you the gold. To win, you must be able to compete each time the baton is passed. IBM won races for a long time, yet eventually Dell defeated the big machine.

Winning the relay race depends not only on the speed and skill of the individual team members, but also on the skillful way in which the baton is passed from runner (supplier) to runner (customers) in the workflow. Compare sprinters from the '80s to those of the new millennium. Today, they're more fit and endure longer, harder, and more precise competition. To be competitive, they have to be Lean. While Compaq remained highly competitive, the belief is IBM fell from first place due to the amount of waste, or fat, in the process. Dell, being a Lean-running machine founded by Mr. Dell in his home by assembling PCs to the customer's requirement without carrying excessive inventory, was able to deliver the product faster than Big Blue. Subsequently, Dell Computer leapt forward in the race through the vision Michael Dell had for the future state of the personal computer market, expanding his company by meeting customer demand for shortened lead times.

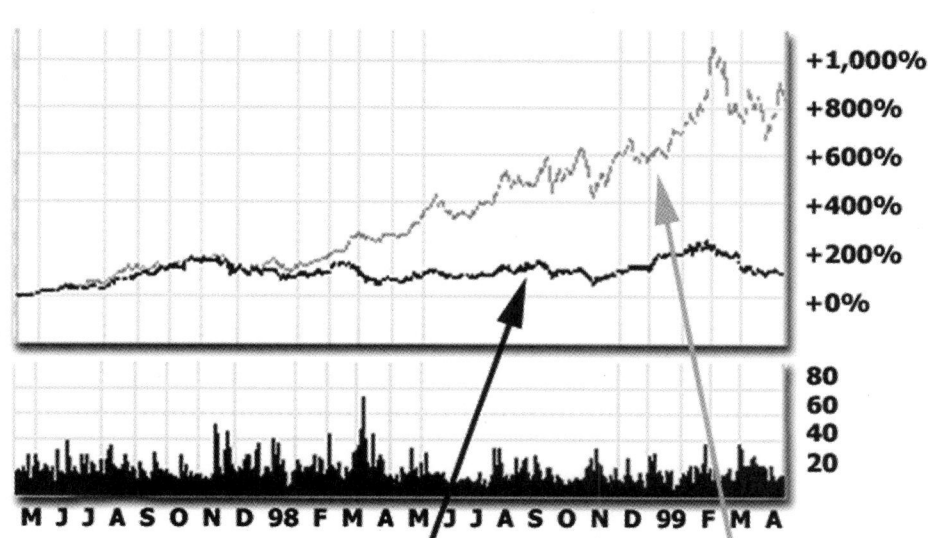

**Other computer companies growth compared to Dell's growth.
Dell's supply chain management value stream is world-class. Suppliers
deliver daily to the customer demand – no more – no less.**

Many companies have trouble simply delivering products without taking speed into consideration. These companies are plagued with exceptionally large amounts of waste in the manufacturing and business operations. When the U.S. Armed Forces needed long johns for the Gulf War, troops were sent to discount stores because there were no U.S. suppliers able to meet the demand without excessive paperwork and processing time. There was no time to wait for suppliers to catch up. The war was on. For the troops, it was a matter of being prepared or facing catastrophe due to the drop in desert temperatures at night.

It's the same in business today. The customer-driven demand for speed, with no excuses for anything less than on-time delivery and perfect quality, has caused a focus on Cycle Time Reduction: streamlining the entire process from the time the customer orders to the time the customer gets his or her canceled check. Today's managers have a heightened interest in the concept of Lean and Cycle Time Reduction and an even greater need to understand the customer's value stream in order to achieve greater shareholder success. You may be asking yourself some questions as you read this book, such as:

Q. I've hired some of the best leaders, managers, and people in my industry. Why can't I just let them use their proven skills to do their job?

A. It's not just about the skills of your individual employees. It's about their interactions and interrelationships. It's about the workflow value stream processes as a whole.

This is where most books on the market get it wrong. Many people look to the teachings of Mr. Shingo and other pioneers of JIT/Kaizen/Reengineering/ Lean philosophy, but something is lacking. There is a growing population of voices speaking out on Lean and looking for a faster, more sound approach to improve workflow processes so their companies can surpass the competition. Michael Dell, although not coming out of an Ivy League school, showed the world how to build computers faster and thus created one the finest supply chain systems worldwide.

A misunderstanding of Lean thinking and its implementation leads to a faulty perception of the requirements for success and, therefore, to ineffective management. This causes minimized results and/or failure to sustain performance for continuous improvement. These companies win one or two legs of the race, but they fail to continue passing the baton to keep on winning. Companies that can leverage the reduction of cycle time can re-invest in the future. So if you think I have just come up with something new about Cycle Time Reduction, you're wrong.

Review the graph below; it shows I have been successful in getting companies to reduce their cycle times.

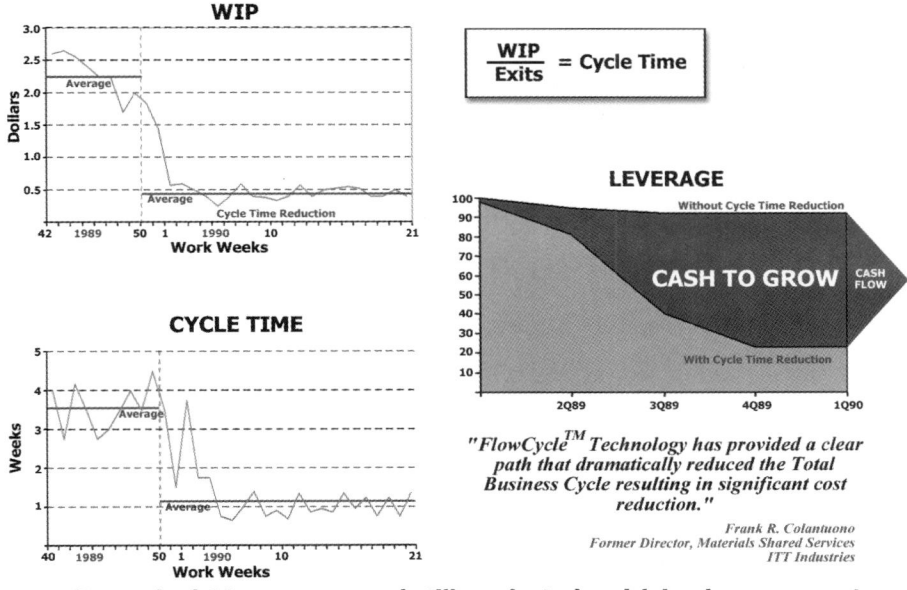

(It worked 10 years ago and still works today driving improvement to cash flow.)

It's not about simply lowering the process times of manufacturing and/or business processes. It's about creating one vision within the organizational culture. This vision makes the whole organization aware of the cost of time and inspires changes every day, if needed, in order to be competitive. In the world today, every process outside of on-time and complete customer deliveries is inconsequential to winning the race.

So what can a business do to keep up with all these demands? The purpose of FlowCycle's type of Lean thinking uses an enterprise-wide approach to maximize the workflow processes. The approach promotes speed and affects the value of the bottom line. FlowCycle is a proven methodology that can achieve needed change in eighteen weeks or less, which makes it extremely valuable in light of today's quickly evolving world.

FlowCycle methodology streamlines Lean into a series of phases that give a clear understanding of the value stream, cut out the waste, and reduce real cycle time. Even more important, these phases are simple enough so they can be applied to every workflow process in your business, not merely the manufacturing of a product. FlowCycle combines culture, business, and manufacturing processes. It is important to recognize that competitiveness through a leaner workflow will result in faster responsiveness to internal and external customers.

FlowCycle's primary vision is for every company to stay competitive. The leadership of the organization gets the workflow processes right the first time and communicates that vision so the entire team does the right thing every time. Once FlowCycle is implemented, its four phases repeat over and over again to keep winning the race each time the baton is passed. The baton is passed from customer to business to manufacturing to delivery across the finish line, on time, every time. Now there is no need for complex fixes that work for only one situation then require even more fixes to keep muda, or waste, out of the value stream. My belief is this – keep it simple.

Simplicity and rhythm are the basic concepts, as illustrated by the FlowCycle symbol below.

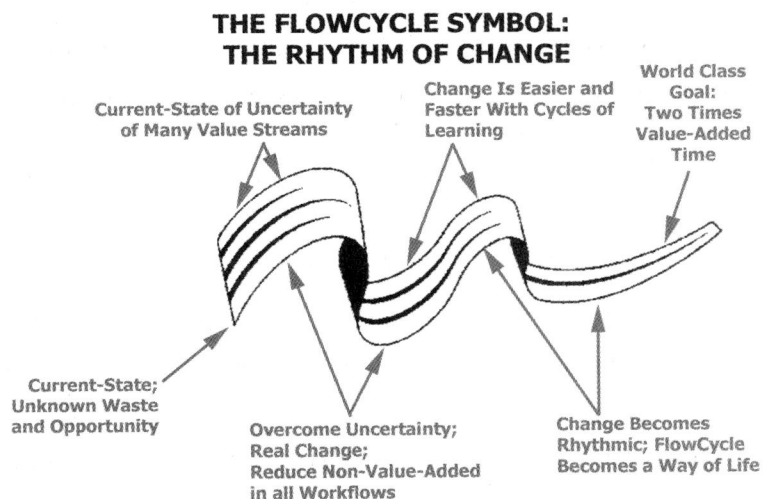

**THE FLOWCYCLE SYMBOL:
THE RHYTHM OF CHANGE**

Current-State of Uncertainty
of Many Value Streams

Change Is Easier and
Faster With Cycles of
Learning

World Class
Goal:
Two Times
Value-Added
Time

Current-State;
Unknown Waste
and Opportunity

Overcome Uncertainty;
Real Change;
Reduce Non-Value-Added
in all Workflows

Change Becomes
Rhythmic; FlowCycle
Becomes a Way of Life

You may not fully understand the meaning of this symbol right now, but keep reading and soon you will learn how it represents changes you need to make to achieve World-Class today as well as changes that will let your business keep winning the gold tomorrow.

CHAPTER ONE

A WASTE-LESS ENTERPRISE™ OR LEAN MANUFACTURING?

FlowCycle Is the Enabler for World-Class Competitiveness

In the new millennium, the management of workflow processes of value streams must be the focus, as opposed to operational thinking or organizational culture. Every process has a flow with non-value-added waste. This is the key principle ensuring workflow cycle efficiency and the fastest way to become Lean, achieve speed to market, and become a World-Class competitor in manufacturing or business process workflows. First place is winning the gold medal. Coming in second or third has less value to both the company and its customers.

Speed is achieved in today's world by trimming waste in enterprise processes – doing away with the non-value-added attitudes and activities. Take track and field athletes as an example: in a relay race, they appear to be Leaner, not just in muscle mass, but also in understanding that clothing is part of the value stream. One goal of the modern track athlete is to create the least amount of wind resistance when running. This goal has changed the requirements of the athlete's clothing to eliminate wind resistance, or waste, in the process.

Loose-fitting clothing, an example of resistance causing waste, was eliminated, taking seconds off the athletes' time. Seconds reduced equals a gold medal. More seconds reduced equals repeatability of the gold medal.

These same principles can be applied to today's businesses. A tremendous shift toward the concept of Lean has occurred in recent years. The new supply chain management era, driven by new technologies, is creating a desire to reduce inventories and push inventories back into the supply chain to shrink space requirements and the caring cost of inventory. It has become more important than ever to reduce the cost of overhead, stabilize erratic production schedule performances, and put further emphasis on the need and desirability of reducing cycle time throughout the entire enterprise value stream.

As new business strategy books hit the market, many company leaders begin reading as much as they can about change. In an effort to stay abreast of the latest strategies, trends, and philosophies, many leaders seek out advice and input from their peers. They often invest in the guidance of outside coaches or consultants in search of the latest improvement tools.

Many well-meaning executives buy into "cookie cutter" plans derived from Lean manufacturing without realizing the total impact these plans can have on the entire business. Unfortunately, once they board this bandwagon, many of these organizations learn the price they pay is much greater than anticipated – and not only monetarily. These companies fail to realize the tremendous benefits that should occur on the enterprise value stream. They get locked into following Lean Manufacturing versus Lean Thinking. As Jim Womack, author of the book *Lean Thinking*, observed:

> Business leaders throughout the world agree on at least one thing: the new millennium will bring new challenges and relentless, continuous change. Globalization, the increasing intensity of competition and new technology, as well as changing managing behaviors and values combine to create an environment of permanent complexity in all sectors of the enterprise within the world.

So what are the alternatives? Is there any merit to FlowCycle, a true Lean Thinking Methodology? What makes it so revolutionary in comparison to the concept of Lean Manufacturing and other methodologies? Let's take a look.

FlowCycle – Managing Flow Every Day

What is FlowCycle? How did it come about? Why does it quickly produce dramatic results?

Let's start with what it is not:

- FlowCycle is not another program of the month.
- FlowCycle is not merely a collection of Lean Manufacturing toolsets and techniques such as Kanban/Pull Systems, Setup Reduction, Defect Prevention, 5-S Program, Team Training, Value Stream Mapping, and other process improvement tools available.
- FlowCycle is not about applying principles of the Toyota Production System (TPS), Just-In-Time (JIT) inventory methods, Total Quality Management (TQM), Reengineering and Kaizen Events.
- FlowCycle is more than Six Sigma.

Six Sigma is a defect prevention toolset developed by Motorola. The goal of Six Sigma is to reduce defects in manufactured products to infinitesimal percentages. Six Sigma training is an excellent tool as long as it is kept in the proper context of meeting customer requirements. A company can Six Sigma itself into oblivion by engaging in quality overkill – taking quality to a point not required by the customer and therefore not valuable to the customer. Managers should remember Six Sigma is one tool to apply within the context of a comprehensive, continuous improvement journey. Six Sigma fits neatly into the FlowCycle execution methodology as a Redesign toolset.

What is FlowCycle?

- FlowCycle is a multidimensional change that begins and ends with people.
- FlowCycle is a way of moving beyond Lean Manufacturing and even

Lean Waste-Less Enterprise by focusing every aspect of the enterprise on achieving speed in meeting customer requirements.

- FlowCycle is an organized, systematic methodology designed for executing change.
- FlowCycle is a daily journey with sustainability. It keeps on working instead of just pulling one or two tricks out of a hat to meet the monthly, quarterly and yearly end financial numbers.

Having a Lean Enterprise does not guarantee future competitiveness in the twenty-first century. To remain competitive, enterprises must manage flow in business, manufacturing and service processes every day. Accounting processes are in desperate need of changing to new and more flexible practices. Being Lean today does not ensure Lean tomorrow. FlowCycle reaches into every aspect of an enterprise, providing an execution strategy for quick response to opportunities and issues as they arise. Having Lean Manufacturing processes is necessary, but it is only one component in a manufacturing plant. Examples and success stories of Toyota Production System (TPS), Just-In-Time (JIT) inventory control methods, Total Quality Management (TQM), Reengineering and Kaizen Events processes improvements abound.

However, these enterprises may still lose dollars to waste in the business processes. It is possible to have the leanest imaginable manufacturing processes and still fail to win the gold.

While preparing to speak to the management of a communications company in Texas, I asked for a tour of a PC board manufacturing line that had won several national quality awards. As we toured the assembly line, I made a quick calculation of the overall efficiency of the manufacturing process. I divided the value-added time down the longest path by the observed cycle time. The line was 98.2 percent non-value-added! It was less than 2 percent workflow cycle efficient. There was a 98 percent opportunity for further improvement.

As I spoke to the management team, I wondered why they were bragging about this particular assembly line when there was so much room for improvement. Even though experts recognized this assembly line as a world-class example of JIT/Lean Manufacturing, only one and a half years later this particular business fell to the competition. Competition overtook them, and the communications company sold that business.

TPS/Lean systems gave Japanese, and subsequently U.S., manufacturers reduced costs and greater flexibility in their production processes. The TPS/Lean system of production enabled managers to eliminate numerous forms of waste, reducing cycle times throughout the value stream. However, TPS/Lean, like mass production systems, remains limited by its focus on manufacturing and the elimination of particular production constraints. Many TPS/Lean Manufacturing programs essentially limit plants to what are called Kaizen Events and improve existing production processes of one product family and nothing more. Many recognize these limitations as a major impediment to achieving even greater reductions of waste and increasing the capacity for utilization of all processes.

While it is true many companies are making great strides in their manufacturing processes, it is also true that many companies are creating the illusion of Lean.

These companies are engaging in large quantities of activities in the form of training programs, Kaizen Events and software purchases, but actual comprehensive financial benefits are still beyond the horizon. Consider this statement made by a professional golfer: "If your best shots are your practice swings and the 'gimme' putt, you might want to reconsider playing golf." (Anonymous quote from the book *Get a Grip!* by John M. Capozzi.) The same principle applies to business.

If your best measure of Lean is the number of people who receive training, the number of employees downsized, or the millions spent on software, you are not implementing the true concepts of Lean.

There has been an ongoing revolution in major U.S. corporations for years. Since Peters and Waterman's watershed book, *In Search of Excellence*, due to the importation of the Toyota Production System from Japan and the Deming quality system in the early 1980s, organizations have been experimenting with ways to increase employee involvement. Some organizations succeed, others fail, but these companies' managers and employees are providing living textbooks that can map the way to new organization models.

The new model is the organization that treats people as its most valuable asset.

Q. Isn't it enough to have many employee initiatives at my company?

A. If you are just getting input from employees and throwing some people on a team, no! Instead, you should be training your people with new skills to visually see waste and remove it for good.

FlowCycle begins with people and ends with people. FlowCycle is a philosophy for managing and viewing the flow of all processes in the value stream. It is not only effective today, but will continue to be effective tomorrow and into the future.

Why? Because FlowCycle companies believe their employees are the solution, not the problem.

FlowCycle is a World-Class concept emphasizing the quality of the enterprise's value stream. FlowCycle defines quality in the three key areas: Total Quality Culture, Business Quality Processes, and Continuous Quality Manufacturing. It is the quality of an enterprise's value stream that will cause it to surge ahead of the competition.

Reaching new levels of quality will only occur if employee teams eliminate non-value-added activities throughout the value stream in the business processes, the factory, suppliers, and even service processes such as law firms and restaurants. FlowCycle allows companies to focus on the basics of managing workflow processes every day to achieve and maintain success. FlowCycle emphasizes people, elimination of waste, and total quality.

Many corporate executives no longer consider TPS/Lean production good enough to maintain a competitive edge. They believe it does not, and cannot, always enable a corporation to meet customer performance demands and unanticipated customer requirements or capitalize quickly on new market opportunities. A TPS/Lean system of production is not fully capable of functioning effectively in the new millennium in which time is the competitive weapon and the competitive focus of business activity is innovation, not simply cost of production.

FlowCycle is a workflow, value-stream-mapping paradigm (a way of viewing the world we live in) that enables companies to reinvent themselves by having every employee involved every day in managing the time of flow and reducing non-value-added time. We have seen the successes and the failures of many dot-com companies and the rise and fall of many well known companies. Every day, companies are reinventing themselves. IBM may have been a leader in the PC and mainframe market at one time, but it reached deep to understand it would have to hire a new CEO from outside to regain market share by going back to its core competencies.

The Volkswagen Beetle is another example of reinvention. The Bug has reinvented itself to meet the customer's needs in a small car and captured a market segment that found it fashionable. It has the highest cost per pound of any car on the market today, yet when the "new" version of the Bug first came out, customers had to wait in line to buy one – and they were willing to do so. Volkswagen did not go back to its old ways of mass production. It found customers who were willing to wait because Volkswagen met its commitments with 99 percent on-time delivery. It did not go back to the batch mentality that created warehouses of surplus inventory for years after marketing the first Beetle.

Repackaging a product for success

Volkswagen gave its Beetle a makeover, and its image made a U-turn.

FlowCycle is synonymous with *change of workflow efficiency*. It is designed to empower employees by giving them much greater visibility, flexibility, and accountability than other systems of improvement. FlowCycle enables employees to clearly see and meet daily customer demand. Proponents of Lean fully appreciate that manufacturing evolves when each system of production builds on past practices. Consequently, tactics such as TPS, Demand Flow, Kaizen Events, Total Quality Management, Reengineering, and Lean Manufacturing are milestones in the journey to Waste-Less Enterprise.

FlowCycle builds upon the TPS/Lean "achievements" in terms of streamlining the workflow processes and continuously improving quality while overcoming limitations in order to make every aspect of a business waste-free.

This means that leaders implementing FlowCycle embrace TPS/Lean system production concepts, but take it farther. Understanding TPS/Lean is an essential prerequisite or foundation for launching FlowCycle. At that point, a company can shift its focus from simple cost-cutting toward increased market growth and create new value streams ahead of the competition. FlowCycle organizations have flexible, dynamic leadership able to shift course rapidly because of their ability to focus across workflow value streams rather than within them.

FlowCycle is a holistic approach to improvement and surpasses the limited capabilities of the JIT/Lean system of production because of its unprecedented ability to make enterprise-wide change every day, even when there is little advance notice in situations such as changes in customer expectations.

> The first priority of FlowCycle is to provide for, and
> continuously ensure, customer satisfaction externally
> and then internally.

We should never forget the simple fact that we are all customers and suppliers in the value stream at one time or another.

What Is a Workflow?

Every workflow process has a flow.™ Processes include manufacturing processes, business processes, service processes, and even personal processes. Workflow is the series of processes, activities, and tasks completed to meet customer requirements. The flow involves repeating processes, activities, and tasks in the workflow to meet customer requirements.

Preparing for work every morning is an example of a personal process. A person might wake to an alarm, stare at the clock in disgust, look at his or her spouse, smile, roll out of bed, and head to the bathroom. He or she might stumble into the kitchen, turn on the coffee, trudge out the front door to grab the newspaper, holler up the staircase to his or her teenage daughter or son, "Time to get up," and so on.

This series of tasks is repeated again and again, day after day. This is the cycle. The above example is a daily cycle. A person's value stream is the entire collection of processes that are regularly repeated.

The goal of any business leader, and consequently the business, is to create wealth by fulfilling customer requirements.

> Any business leader wants his or her business to expend the least amount of effort necessary to achieve the desired result of meeting customer requirements.

Expending the least amount of effort means making the minimum capital investment and operating the business while expending the smallest amount of resources possible. Through this, shareholders in a business maximize the return on investment. Expending more effort than required is wasteful and reduces returns.

For example, a newspaper delivery person does not throw a newspaper to the house on the corner, then drive around the block before delivering the newspaper to the house next door. The person throws papers to each house in sequence. A professional bass fisherman does not cast a lure fifty feet into the air to make a six-foot cast. The fisherman flips the bait directly to the desired spot.

One would think there would be no need to explain such simple concepts, but take a walk through most businesses and factories, and you may begin to wonder. Try this exercise: pick anyone you work with and ask them to describe, in sequence, the steps they go through to accomplish some simple daily activities. You will go away scratching your head.

> At a Texas factory, FlowCycle implementation employee teams reduced a workflow process where their product traveled over six miles down to two hundred feet! How could a process get that bad and stay that way for years in a business employing many very talented and intelligent people?

Amazingly, situations like the one above are the norm, not the exception. That is why companies must teach basic, Cycle Time Reduction concepts and why Cycle Time Reduction thinking must be the norm, not the exception.

How Workflows Fit in the Value Stream

In a business, the value stream is the entire series of workflow processes from the headwaters all the way downstream until the customer drops cash permanently in the company's bank account as a result of meeting requirements. A series of interconnected workflow processes comprise the value stream. Within any of the processes are series of interdependent activities that people or machines accomplish by completing a series of tasks.

The graphic below illustrates the types of waste in workflow process that consume time and waste financial resources.

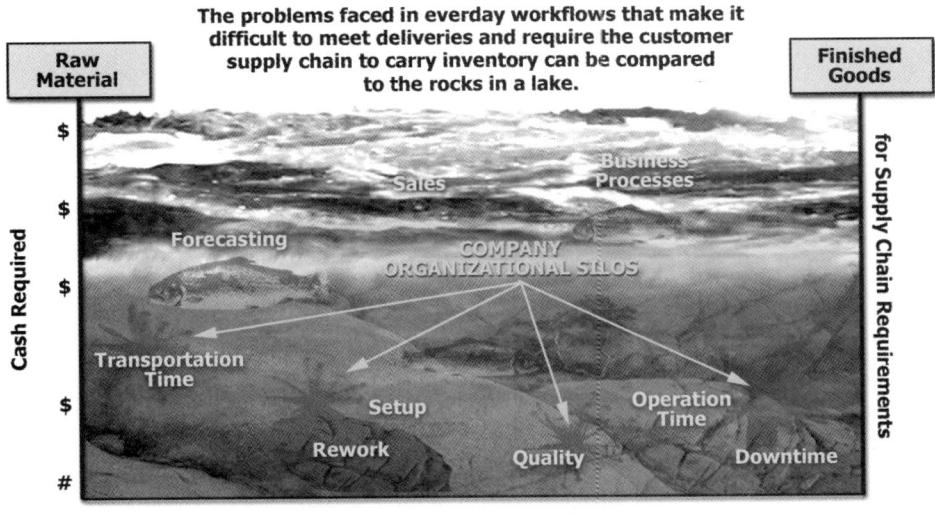

Large inventories and long cycle times cover up waste in processes and imprison precious financial resources.

The water level represents cash needed to run the business. When your team can increase flow from the headwaters to spill over the dam at a faster rate, the water level will drop with less resistance. Throughout this book, the terms value stream of Workflow Processes equal the Activities and Tasks that slow the flow of water being pulled through the dam to meet customer requirements. Workflow has to be managed to increase the speed of the flow.

Improving Flow

If the workflow processes that comprise your business value stream wander around like the Israelites in the wilderness before finally entering the promised land of meeting customer requirements, you can be sure of two things. First, customers no longer wait forty years for satisfaction, much less forty days, forty hours or forty nanoseconds in some cases. Second, competitors are intensely studying ways of providing a better product or service to your customer at a lower price and faster than you.

As businesses enter the twenty-first century, they face two issues whose intensity seems to be increasing exponentially: demanding, disloyal customers and relentless competitors that seem to be appearing from out of nowhere on every horizon.

The American automobile industry came under attack from foreign competitors in the '70s from a company who improved its processes. In the early '70s, when Toyota began gaining market share, they developed a

"Just-In-Time" production strategy to respond to customer demand on a daily basis, carrying no more and no less inventory than required to meet that demand. Toyota's JIT strategy became a flexible, disciplined approach with a common set of goals defined by three simple principles: to process, produce, and deliver products consistently, exceeding expectations in terms of quality, cost, and time. JIT built upon a manufacturing philosophy that shortened the lead time between the customer order and delivery through the elimination of waste in workflow processes. Toyota created the Toyota Production System to gain competitive advantage by improving workflow processes. Managing flow to improve processes is nothing new.

One-Piece Flow

"One-Piece Flow" is a classic manufacturing technique that demonstrates attempts at improving flow. The ancient Egyptians used one-piece flow to build the pyramids. The assembly line was a long string of people moving one block at a time at a rate that met the demand of placing each block in its final destination. Any interruption in the flow of the blocks toward the final destination would cause problems in every step, before and after. The goal is to keep the flow moving. Once the flow slows or stops, waste begins to accumulate in the form of scrap, rework, downtime, inventory, and late shipments. The rate of flow to meet customer demand is a concept known as "Takt rate," which will be described later.

In the early 1900s, Henry Ford used the concept of one-piece flow to assemble autos affordable to all. The concept was to balance operations based on time, smoothing flow, and keeping the flow moving. In theory, one-piece flow production is the perfect way to manufacture a product, but life is complicated, and so are products.

The success of the Model T was one model, one color. This was also the failure of Ford, at one time.

The concept was to balance each operation to create an even flow based on time.

Ford found this out when a lack of flexibility ultimately cost him his market share. Customers were willing to pay for options such as a variety of colors. Ford failed to understand the true value of his product. Chevrolet better understood the value of the product and offered customers a choice of colors. Since it took time to mix paints, lead times increased, creating other bottlenecks and constraints in the production processes. Customers perceived a choice of colors as valuable. At first, customers were willing to wait for optional items, but variety kept stretching the order fulfillment leadtime, ballooning inventories, increasing costs, decreasing quality, and eventually, alienating customers.

FlowCycle – An Execution Strategy to Marry One-Piece Flow and Variety

TPS (Toyota Production System) and Lean Manufacturing methods attempt to marry one-piece flow with product variations as a means to overcome the limitations of one-piece flow and the problems created by offering customers optional equipment. Implementing JIT tools is not easy while a factory continues to produce a product every day. Due to a lack of market share in the early 1970s, Toyota was able to stop its factory processes and take the time necessary to eliminate waste without hurting ongoing customer relationships. This allowed Toyota to raise quality and reduce prices, leading to increased market share, just as Henry Ford did. This graphic illustrates what Toyota did.

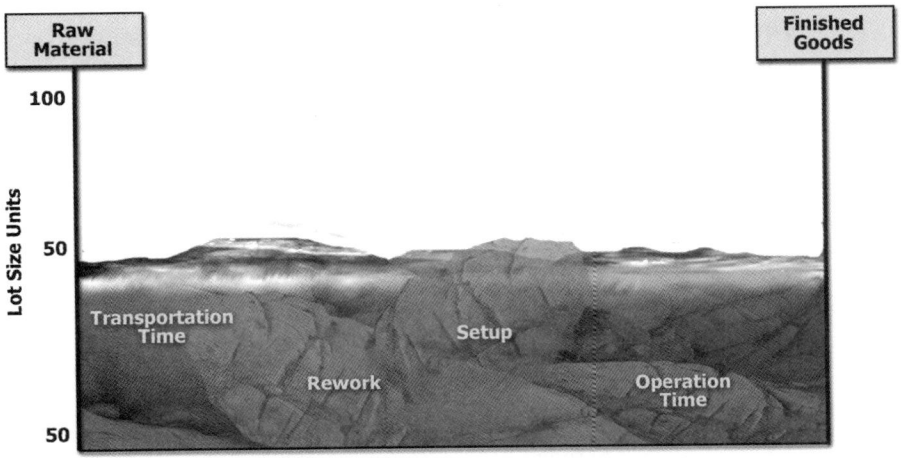

Traditional JIT used trial and error batch size reductions: very slow and painful!!

Toyota shut down production and drained inventory levels to the point where major constraints became easily identifiable. A central principle of Lean is that inventory hides problems. Some companies allow inventory levels to increase so that they can overcome problems upstream from their process. Lowering inventory without a corresponding workflow improvement strategy is dangerous to customer satisfaction. Toyota realized that to make the required improvements quickly, they had to suffer short-term pain and risk alienating current customers. The goal, of course, was to earn many more new customers with higher-quality, lower-cost vehicles. The strategy was a spectacular success for Toyota, but not one it could repeat in today's competitive market.

Very few companies have the luxury to cease operations while they redesign processes and implement improvements. This is why businesses require an execution methodology that leaders can imbed in a company culture very quickly, applying process improvement techniques and improving customer relationships continuously.

FlowCycle – The MARI Execution Methodology

It does not matter where a company stands in its workflow process improvement journey. *FlowCycle – The MARI Execution Methodology* starts where an organization currently stands and builds on what already exists to methodically reduce waste in all workflow processes. The goal of FlowCycle is to eliminate unnecessary activities and tasks (those that are non-value-added in meeting customer requirements) to save time, reduce effort and costs, and increase cash flow. Ultimately what remains are those activities necessary to meet customer expectations and business requirements.

The Number-One Constraint

Without question, the single greatest factor affecting the efficiency of workflow processes is the behaviors and attitudes of the people involved in that workflow. To become a World-Class competitor, a business must turn the number-one flow constraint into the number one flow facilitator by empowering employees to focus on Cycle Time Reduction in their processes every day.

This is easily accomplished by:

- Having Human Resources develop company culture around total customer satisfaction (customers, both internal and external), making it the number-one goal of every employee.
- Proactively communicating all requirements to all employees.
- Providing problem-solving training, tools, and education opportunities that facilitate repeatable cycles of learning to drive waste out.
- Implementing new accounting measures to drive true behavior of the management by eliminating traditional cost methods to Lean/Simple measures..

As Johnny Miller, twelve-time winner on the PGA tour and two-time PGA champion, said, "No one becomes a champion without help."

Successful companies keep the focus simple on this key principle: Companies should meet or exceed customer expectations on a continuous basis by expertly managing the workflow processes comprising its value stream.

To maintain and increase market share, leaders must focus significant resources on process improvements. Without daily exercises in waste elimination, workflow processes will become fat, and lazy and wander aimlessly, consuming time.

FlowCycle – the MARI Execution Methodology, is a leadership strategy and philosophy designed to keep continuous improvement, and Cycle Time Reduction principles as core, fundamental, recurring patterns in the minds of all employees of an organization.

The FlowCycle Approach to Achieve Continuous Improvement

FlowCycle is successful in literally every company where it has been applied. FlowCycle is a customer and workflow process management philosophy for establishing organized, continuous process improvement in all activities. The concept of this methodology is to build into an organization's culture cycles of learning that can be repeated. Remember, winning one gold medal is not good enough. Winning means continuous victory. FlowCycle is a vision-based strategy that gives a business the execution methodology and tools to accomplish that vision.

FlowCycle provides the motivation to continuously improve. It involves everyone – whether in a line, staff, or support function – at every level of an organization. The outcome is increased customer satisfaction regarding all of an organization's products and services, attained by everyone doing the right things the first time, on time.

FlowCycle is designed as a long-range improvement strategy. To achieve daily progress, planning must be thorough enough to support the established day-to-day goals, but the daily action items imposed for improvement must not be overly complicated. In other words, "keep it simple" so improvements will continue.

To improve a process, the people in the process must first understand the process in its current state. The key to understanding a process is to make it visible in order to make it a call to action.

FlowCycle evaluates value stream process flows in two separate areas with a visual mapping process. Employee teams develop value stream maps from both a high-level business and a detail workflow process perspective. The process maps and MARI provide employee teams a common language that sets up a framework for improvement throughout an organization.

"Change before you have to," says Jack Welch, CEO of General Electric and the most quoted manager in the world today. He talks about "Continuous Improvement" (incremental and evolutionary) change and "transformation" (quantum and revolutionary) change, urging companies to embrace both types as the only way to survive in today's global, competitive environment. Jack sums it up with the following:

"Faster, in almost every case, is better. From decision-making to deal-making to communication to product introduction – speed, more often than not, ends up being the competitive differentiator."

The Measures of World Class to Maintain Continuous Improvement

Which characterizes the ongoing health of your organization – robust improvement, stagnation, or careening downhill? What percentage of activities and tasks in your company's workflow processes are value-added? FlowCycle teaches businesses an overall measure of World Class called "Workflow Cycle Efficiency" and to daily depend on six key visible measures to drive efficiency.

Workflow Cycle Efficiency (WCE), the Measure to Focus on World-Class Speed

Workflow Cycle Efficiency (WCE) is an equation used to demonstrate the overall efficiency of a value stream. The equation is "value-added time down the longest path" divided by the "observed cycle time."

The Football Game Value Stream as an Example of WCE

Calculating the Workflow Cycle Efficiency of football illustrates the concept of value-added, non-value-added, and overall efficiency. In 1957 when the Chicago Bears played the Green Bay Packers, a fan could go from the car into the stadium and watch the football game in 78 minutes. Today, the game process takes 180 minutes to accomplish the same goal. The core design of football has not changed. It remains four fifteen-minute quarters for a total sixty minutes of game time. However, the observed cycle time has increased from 78 minutes to 180 minutes. The objective remains the same – for the offense to execute plays, move the pigskin down the field, and score as many points as possible as fast as possible.

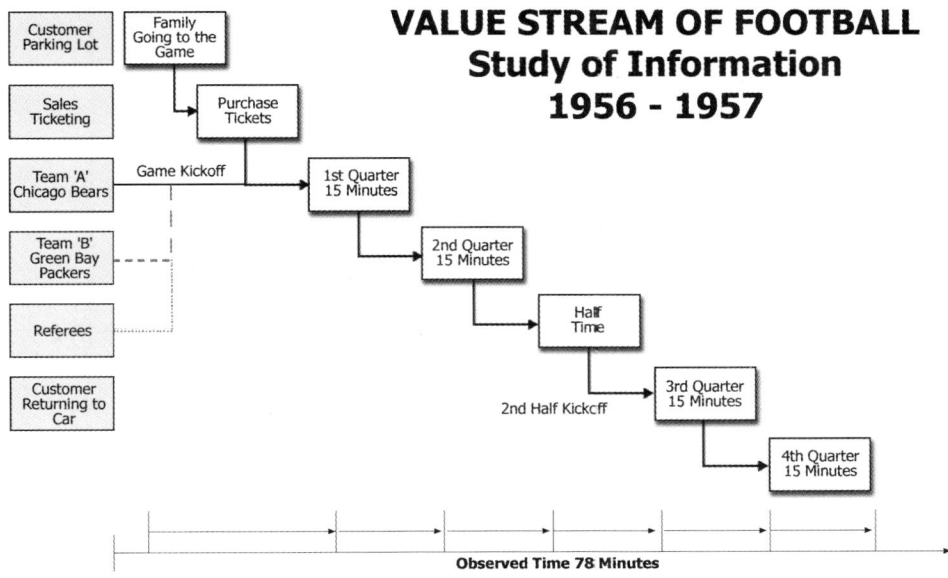

What are the value-added activities or tasks in a football game? The only true value-added time is the time from when the ball is snapped until the referee whistles the play over. The offensive play may be running, passing, or a kick, but when the whistle stops play, value-added time ceases and non-value-added time begins. Television commercials, halftime festivities, walking back to the huddle, arguing with the referee, chatting with the coach, and setting up at the line of scrimmage are all non-value-added activities. None of these activities actually move the ball toward the goal of scoring points, so they are non-value-added. The official time clock runs for sixty minutes, but what is the value-added portion of that sixty minutes? From a study done of 2,000 NFL and college games, teams are in the process of executing plays for eight minutes and fifty-four seconds (see the graphic below).

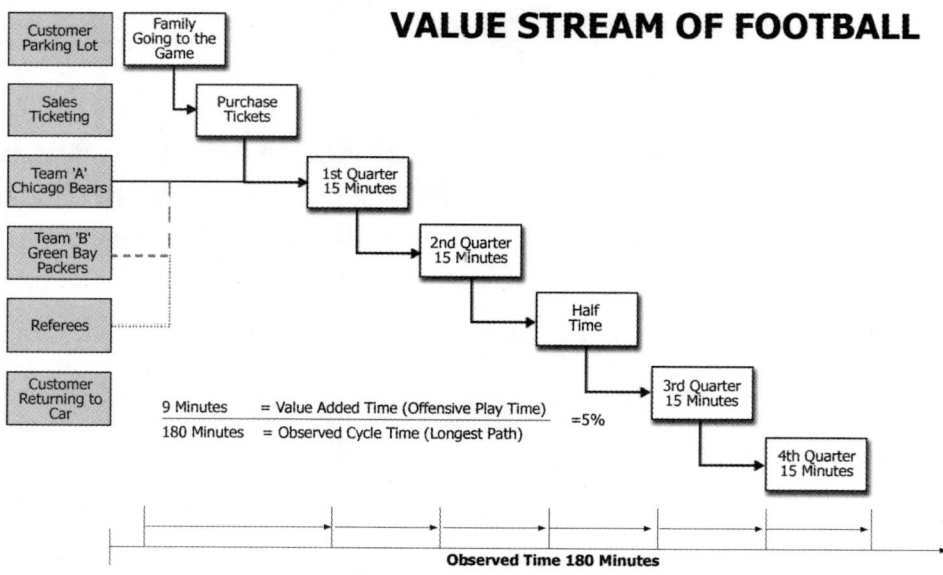

To calculate WCE, take the "value-added time down the longest path" divided by the "observed cycle time." Football's Workflow Cycle Efficiency is about five percent – nine minutes divided by one hundred eighty minutes. Sound bad? One would hope that businesses with the goal of a profit would have a better WCE than football. But in over twenty years of observing workflows, I personally have yet to observe a workflow process that is greater than two percent workflow cycle efficient. In other words, 98 percent of the cycle time is consumed by wasteful, non-value-added activities. The good news is there is a 98 percent opportunity for improvement.

The football value stream also illustrates how non-value-added activities have a tendency to balloon over the years.

Add a little step here and a little step there, and before you know it, the entire company is spending hours a day watching beer commercials, so to speak.

In the case of football, it can be argued that the entire 180 minutes constitutes a value-added entertainment experience for which a customer is willing to pay. That may be true in a sense, but consider this: take away the nine minutes of actual play – would anyone sit through the other 171 minutes? Probably not!

Think about the Workflow Cycle Efficiency of the various processes in your company. How many of those processes, which may have been very good many years ago, have evolved into processes that contain massive amounts of waste? The target for World Class, as described by Toyota's Mr. Shingo, are processes that are fifty percent Workflow Cycle Efficient; that is, half value-added and half non-value-added.

Apply this equation to the best process in your business, and it will help make it clear just how much room for improvement remains. Workflow Cycle Efficiency (WCE) is applicable to both business processes and to the manufacturing sector. Divide the "value-added time down the longest path" by the "observed cycle time" to calculate Workflow Cycle Efficiency. WCE is an enabler to keep organizations focused on continuous improvement and waste elimination.

Accounting Tracking Effort

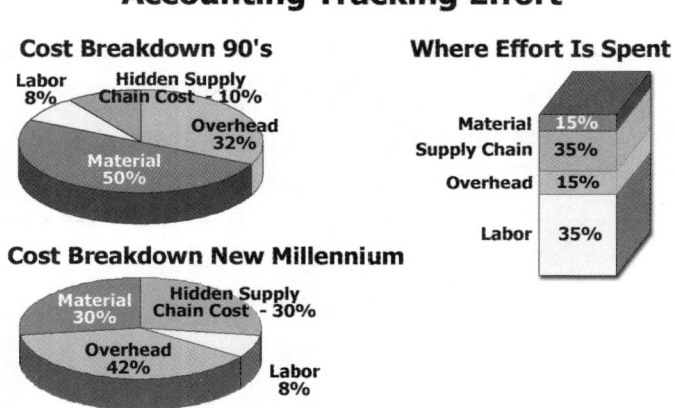

The focus is on supply chain and labor – but I have seen compaines still get in trouble or go out of business. Why? Because they don't maintain the following six measures that capture time and all costs associated to running a business.

Six Key Measures That Drive Improvement to Achieve a Higher WCE

To support WCE improvement, there are six key measures that business, service, and manufacturing improvement teams should monitor for improving a company's value equations. These charts are graphed monthly, weekly, and daily for accountability to create a sustainable visual process.

1. *Conformance to the customer schedule:* on time and complete to scheduled date. This is the most important measure. To create a true pull system, it is important to be on time and complete with internal customers in order to meet external deadlines. Teams should not be allowed to use the customer as an excuse for missing a customer demand or failing to improve.

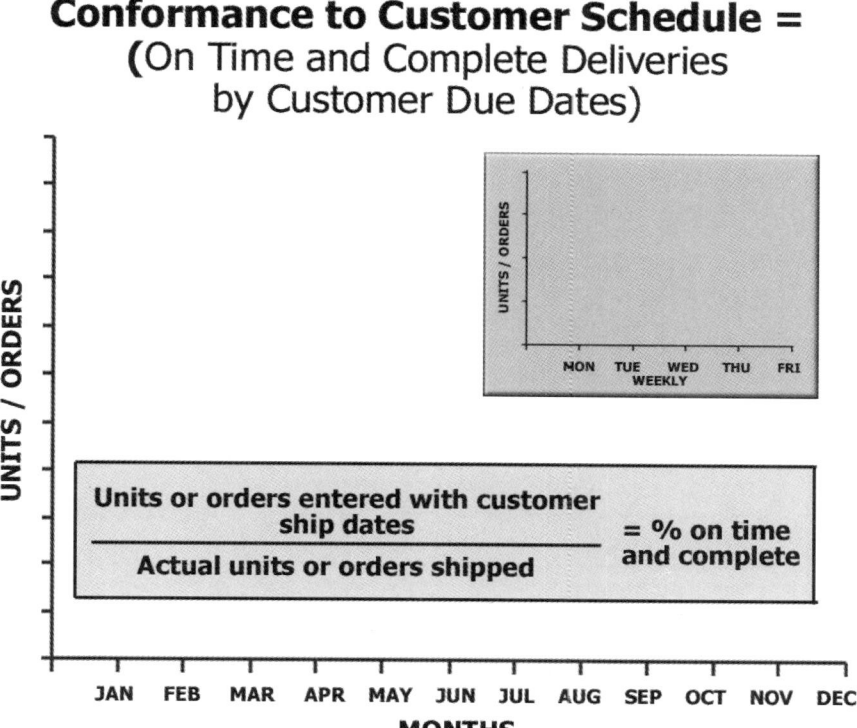

Conformance to Customer Schedule =
(On Time and Complete Deliveries by Customer Due Dates)

$$\frac{\text{Units or orders entered with customer ship dates}}{\text{Actual units or orders shipped}} = \text{\% on time and complete}$$

2. *Cycle time* of a manufacturing (such as an assembly line) or a business (such as engineering design, accounts payable or human resources) process. This measure forces teams to focus on continually reducing cycle time in every workflow. Cycle time is calculated by taking work in process (WIP) divided by the actual exits/shipments in a scheduled tracking period.

Cycle Time Measure
(Cycle Time Reduction)

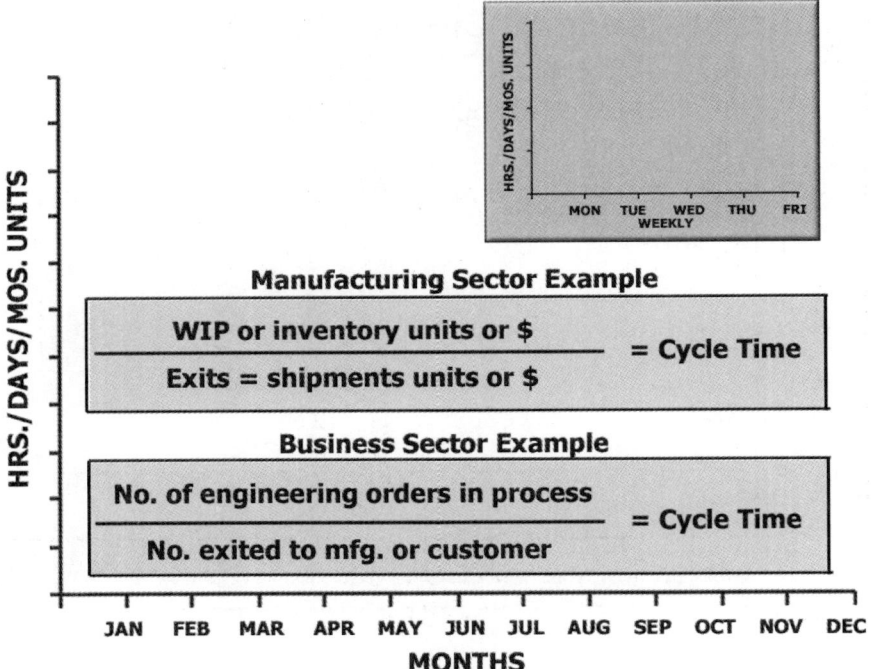

Manufacturing Sector Example

$$\frac{\text{WIP or inventory units or \$}}{\text{Exits = shipments units or \$}} = \text{Cycle Time}$$

Business Sector Example

$$\frac{\text{No. of engineering orders in process}}{\text{No. exited to mfg. or customer}} = \text{Cycle Time}$$

Cycle time is defined as the duration required to satisfy customer requirements from the point a need is expressed to the moment the need is satisfied. Cycle time includes lost time due to delays in the process due to nonconformance. With the lake full, as seen in the earlier graph, how long the flow will take from the headwaters to the dam is the answer a customer will want to know.

36

3. *Inventory:* total or work in process. This is a total inventory of raw material, work in process, and finished goods. Inventories in business processes are measurable also.

Inventory Reduction

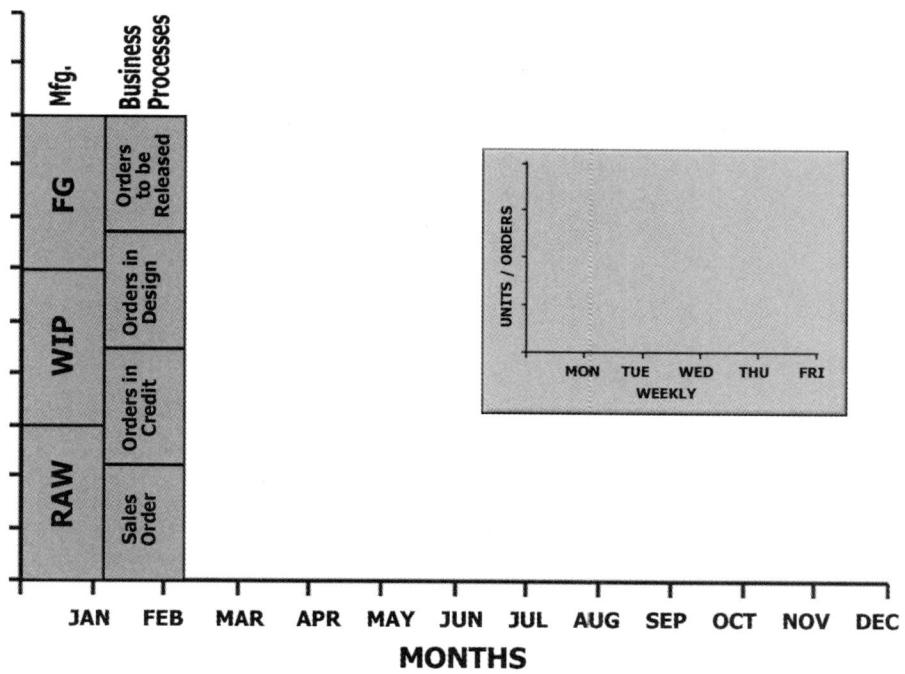

4. *Takt time:* the pace of the process to meet customer requirements. The Takt rate of the business or factory process is the pace of the flow to feed the customer product, information or services at the rate the customer requires. Takt is a German word meaning the rhythm to monitor the RPMs of an engine to ensure it will not die or stall. Takt is a key measurement to monitor throughput to ensure the downstream operation has the correct fuel or product mix to meet customer demand. Takt rate is key in maintaining Kanban/Pull Systems production.

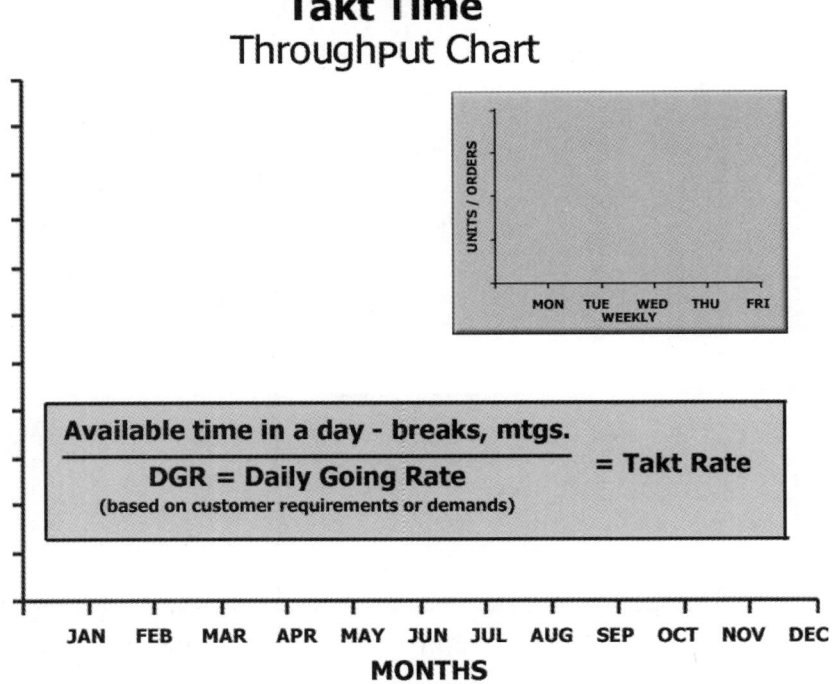

Takt Time
Throughput Chart

UNITS / ORDERS

MON   TUE   WED   THU   FRI
WEEKLY

Available time in a day - breaks, mtgs.
────────────────────────────────────
DGR = Daily Going Rate   **= Takt Rate**
(based on customer requirements or demands)

JAN   FEB   MAR   APR   MAY   JUN   JUL   AUG   SEP   OCT   NOV   DEC
MONTHS

5. *Downtime:* Downtime can be broken down into categories of absenteeism, part shortages, machine downtime, setup, etc., for charting and analysis. Accounting systems focus on tracking many manufacturing variances, but categories of Downtime also exist in business processes. There is a focus on the factory floor cost of labor, but salaried employees often waste time in business practices with no accountability. This waste is Downtime to the internal customer. Downtime measure reveals all lost time. Henry Ford said time is the greatest of all waste; it is just hard to see.

Downtime Chart

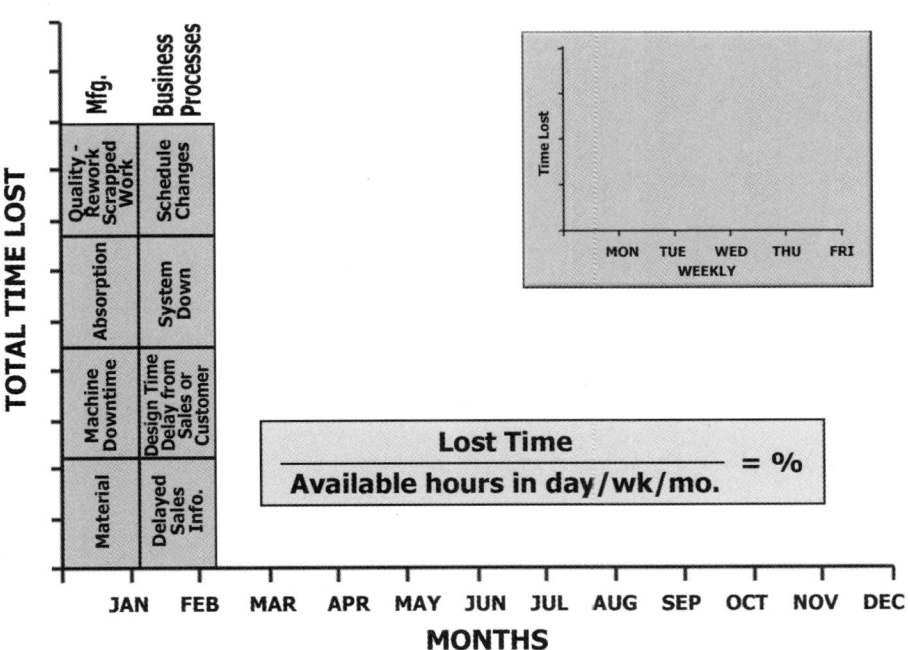

6. *Scrap:* equal to dollars of waste of any kind in manufacturing or business processes. Scrap in the factory is easy to see. Scrap in business processes is more difficult to see; for example, wasted copies, reports, supplies, equipment, or time. For every unit of savings you find in manufacturing processes, you can find six in business processes.

Scrap Dollars Chart

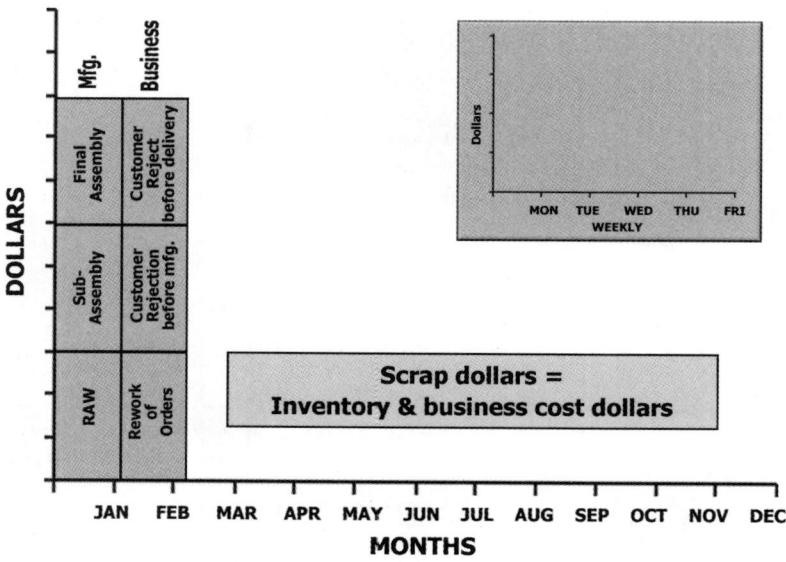

Deploy these six simple measures using visual tools at every level of a business or plant, and they will drive the predictability of improvements.

Employee Motivation and Employee Involvement

I have a primary belief that the basis for continuous improvement can be partially based on the tools contained in the Toyota Production System and on Henry Ford's one-piece flow line.

The product flow was balanced to one-piece flow, but as you can see from this picture, inventory covers up the problems. The bins below the workstations are point of use storage but also increase production cost.

The Toyota Production System is just one of many well-established improvement methodologies. As each company is unique, any improvement initiative must be tailored to each particular situation.

FlowCycle is a framework for solving problems that incorporates the best ideas of others and organizes the execution and implementation of these ideas into process improvements.

FlowCycle's origins can be traced to the early 1980s when IBM sent five industrial engineers to Japan to learn the concepts of Just-In-Time from Toyota's quality guru, Mr. Shingo. Mr. Shingo helped revolutionize manufacturing practices throughout the world.

Upon completing the training, a study was launched to decide how to apply these Japanese techniques. That study yielded a concept IBM called Continuous Flow Manufacturing (CFM).

In spite of their best efforts, results were limited. Why? Because the transfer of JIT/CFM knowledge from a few engineers to the people who had to actually implement the techniques was limited by the number of people that were exposed to these concepts. In other words, there were insufficient numbers of people at any one particular location who understood the concepts and how to apply them. Because of this, successes were isolated, spotty, and inadequate overall.

This is still happening today in the world of Lean Manufacturing. There are many consulting firms selling Kaizen Events as a methodology of improvement. These Events focus on one particular constraining barrier or bottleneck. In a Kaizen Event, an engineer, manager, someone from production control, or maybe a couple of assembly operators work as a team to assess and implement change in a one-week time frame. The goal is to free up an operation that is a constraint and then give that operation back to the workers who implement the changes made by the Kaizen team.

This approach has some fundamental problems. If someone enters your home, rearranges the furniture, repaints the bedroom, and hangs new pictures and you had little or no say in the new design, would you support those changes? Probably not! This illustrates one of the primary reasons Kaizen Events generally fail to produce sustainable results. Kaizen Events often do not involve all employees in all activities of the work processes.

Unfortunately, Kaizen, which means "continuous improvement toward perfection" in Japanese, has been reduced to an "event," which denotes a one-time occurrence.

Based on my experience, once the word Kaizen enters the vocabulary, hope of continuous improvement in the value stream, performance is gone.

*FlowCycle is designed to serve as both a short-range and a long-range improvement journey for implementing fundamental, permanent, positive change. It requires everyone to be committed to the process. **All employees must continually change and learn for change to be continuous.***

Reducing cycle time has a cascading effect on all aspects of cost. As cycle times are reduced, your employee productivity increases proportionally. A fifty percent reduction in workflow cycle efficiency will create double the capacity of your business. As the speed of workflow increases, resource capacity is freed up. Two things happen: costs decline and the organization becomes capable of paying for its improvement process. Once again, the goal is to create a one-to-one ratio, producing significantly more output with the same or fewer resources, a winning combination as long as you sustain this vision of World Class through the view of WCE.

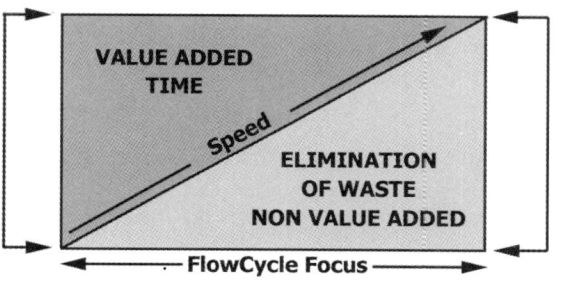

To get to a ratio of 50% Value Added Time, management must allow all employees to always work on elimination of waste.

VALUE ADDED TIME

Speed

ELIMINATION OF WASTE NON VALUE ADDED

FlowCycle Focus

To keep the focus on waste ellimination FlowCycle tools and measurements must always be visible.

The FlowCycle Advantage

Implementing a continuous improvement journey through FlowCycle requires an investment of time, resources, and money, but the rewards can be tremendous. The very act of involving everyone in visibly analyzing workflow processes from the customer's point of view teaches teams to start thinking about streamlining business and manufacturing processes.

A correct business strategy will align the workflow processes with a skilled, learning organization/culture supported by flexible information technology systems.

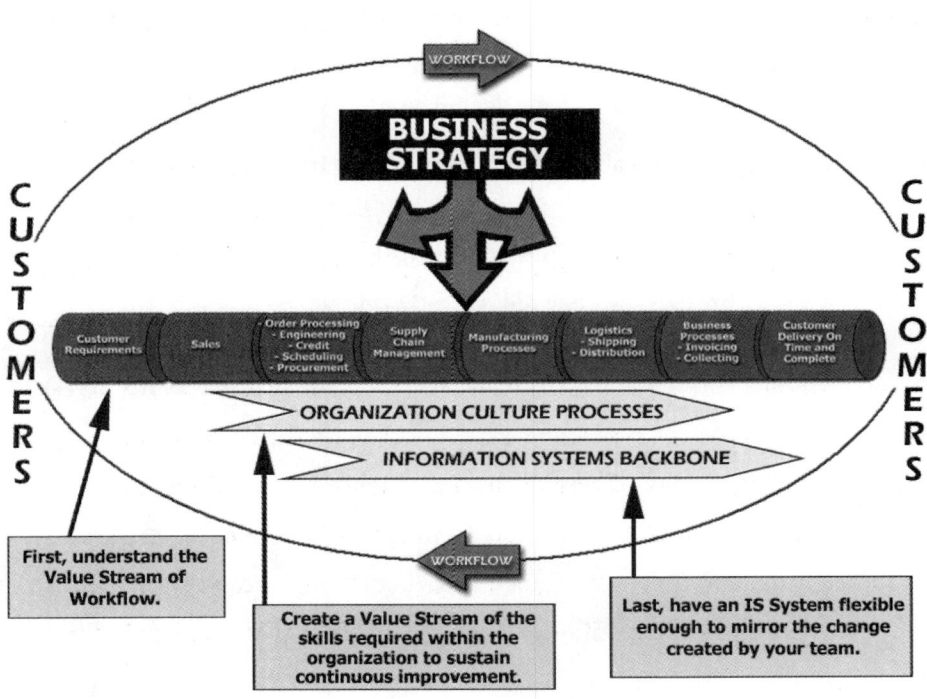

FlowCycle/Lean Workflow Improvement

Lean becomes simple when defined and organized by FlowCycle concepts. Every process has a flow, and that flow can achieve faster cycle times. Within a process flow, there are value-added and non-value-added activities. The goal is not to focus on the value-added time. The goal is to focus on reducing the non-value-added activities to achieve the new definition of World Class, turning the number-one flow constraint into the number-one flow facilitator. This goal keeps the entire organization focused on continually reducing cycle times and becoming leaner.

Let's take Tiger Woods as an example. At the age of 24, Tiger had achieved more than most golfers will achieve in a professional lifetime. When Tiger won the Masters, his first major tournament title, his body fat was just over 11 percent. He did not view winning the Masters as the ultimate achievement. Tiger used MARI to improve. He continued to work with his coaches about mobilizing himself for greater improvement. He continued assessing every aspect of his game. He decided his game was not yet perfect, so he began redesigning it. Finally, he and his coaches began implementing the redesign.

The result was, in the year 2000, he won three consecutive majors and numerous other tournaments. His body fat percentage dropped as his fitness level grew. His redesigned swing and putting stroke reduced variability and increased on-target shots. These efforts ensured his continued success on a worldwide level and increased his lead over competitors. After winning his third major in 2000, Tiger was quoted as saying, "I hope I can continually improve to meet the competition of next year's golfers at the majors." Tiger sets an example that illustrates how we all can continually improve even though we may already be seen as the best.

Tiger's competition is not sitting idly by as he beats them week after week. Tiger's success is causing other golfers of all ages to begin the Mobilization, Assessment, Redesign, and Implementation process on their game. This is the FlowCycle advantage. It provides the methodology (MARI) for actually implementing companywide process improvements with sustainability.

FlowCycle Company Can Outperform the Competition

The FlowCycle execution methodology evolved out of a continuum of personal and business experiences. I found the concepts of Cycle Time Reduction fascinating from early in my career at IBM. My experience in helping develop Continuous Flow Manufacturing (CFM) at IBM and the practical difficulties we faced in actually implementing process changes caused me to really examine what it would take to apply the improvement toolsets to workflow processes.

I left IBM after ten years and spent five years with a start-up consulting firm that grew out of CFM success at IBM in Austin, Texas. Applying CFM tools reduced the cycle time of the PS-2 motherboard from 45 days to 4.5 days in nine months. With the consulting firm, I led and participated in numerous Cycle Time Reduction projects for a diverse group of companies. The need for a holistic execution methodology for dramatic, rapid results continued to come into focus and led to the creation of FlowCycle.

I spent the next two years as an executive at a manufacturing company. At this business, I deployed workflow and Cycle Time Reduction concepts beyond CFM with fantastic results. This company gave me my first opportunity to involve all employees and develop the methodology far beyond what I learned at IBM and the consulting firm.

I began to understand that what organizations need to do is to focus on the quality of their internal culture and view business processes and manufacturing operations as one value chain of activities – that is, as a workflow. It was in this setting I developed an awareness that smaller organizations' ability can outperform large competitors by bringing new products to market very quickly and by responding far more rapidly to the customer's new product requirements.

The company's name was PlyGem Industries. They built Crestline, Vetter, and Great Lakes windows and doors. Its competitors were Anderson, Pella, Peach Tree, and other major window companies. To stay competitive, it was imperative that PlyGem capture market growth by reducing the lead time for the production of custom windows. This was a union environment, and our challenge was getting the employees to understand the need to improve the value stream for customers. We mobilized the culture around a vision of winning based on speed and by showing them the changing market demands that had resulted from America's economic growth.

What happened? We didn't merely reduce cycle time to barely keep up with the competition – we reduced cycle time 90 percent! To their credit, the employees took the average cycle time to build a custom or standard home window down from eight hours to one hour in the production lines. PlyGem was able to respond more quickly to the customer and improve quality at the same time. After PlyGem saw the results of the manufacturing process reduction, they turned to the business processes. We simplified the workflow processes from manufacturing, engineering, order entry, sales, and other areas by mobilizing the culture around the initiative of Cycle Time Reduction for all activities. The Cycle Time Reduction of the business processes reduced lead time from three weeks to one week. The bottom line was the business grew from $176 million to $326 million in two years without adding additional people. The company went from a stock value of $10.24 to $24.00 in less than a year. Value was created for all shareholders in the business.

Unfortunately, this story does not have a happy ending. After one and a half years of improvements with what I called at the time Total Quality Management-III (TQM-III), a new CEO arrived who had different ideas. The different idea was called "Reengineering." This was in 1993. I am not suggesting the concepts of Reengineering being applied in the early 1990s were all bad, but in the case of PlyGem, we had already made enormous improvements in the workflow processes. The new CEO applied Reengineering throughout the company, and in the nine months that followed, he and the consulting firm he hired to help literally destroyed the business. The stock price fell to less than $7.00 a share, and the company was sold off in pieces to a larger conglomerate. What was the fatal error? They did not involve the people.

WARNING: Companies can Reengineer, Just-In-Time, Kaizen Event, and Lean too far. They can stretch people beyond their capabilities or eliminate the wrong activities and in the process destroy company culture, customer satisfaction, and consequently, the long-term health of the business.

The CEOs Conference of 2000 in Washington, D.C. stated that only 39 percent of CEOs were satisfied with their Lean programs. Many corporate executives no longer consider the Lean system of production good enough for a corporation to maintain its competitive position. They believe it does not, and cannot, enable a corporation to effectively meet the performance demands of today's competitive markets.

So what happened? They all had good reasons for Lean thinking. Seventy-nine percent of CEOs said the reason was to reduce cost of the supply chain and move the inventory back to the supply chain, 48 percent said to improve employee satisfaction, 67 percent said to acquire other companies, 89 percent said to increase customer satisfaction, and 74 percent said they were doing Lean to improve their profitability and market share. In fact, these goals should have been attainable, but only 39 percent were satisfied.

A well-designed and applied Lean/FlowCycle journey will have the following consequences:

- Changes to existing jobs duties/structures/skills
- Integration of independent business units in value streams
- Increased authority of process owners to make improvements
- Increased corporate control of vision to increase shareholder value

So why weren't these CEOs able to achieve their goals? There are three top factors preventing greater Lean implementation success:

1. Lack of organizational readiness by leaders and managers
2. Organizational resistance to change due to a failure to change personal performance plans
3. Insufficient executive leadership

The number-one fatal Lean implementation error is a failure by CEOs to invest adequately in change management and communications tools. These tools ensure the implementation achieves a culture transformation, which is a vital prerequisite to successful application of any change initiative.

To quote a Deming phrase, "No Theory, No Learning." Deming warns that people must fully understand the theory of a Waste-Less Enterprise before they will embrace learning and applying the detail tasks required to achieve it.

Employees must be free to examine and discard assumptions based on past beliefs for performance breakthroughs to occur. Unless someone or something challenges fundamental beliefs, theories, assumptions, and thinking for validity, the future state will rest on the same foundation as the past and will be essentially unchanged. The result is employees who are forever inventing different versions of what they have always done in the past. No real change will occur, just different manifestations of the same old thing.

In the twenty-first century, with customers demanding results and competitors fine-tuning their engines, business leaders do not have the luxury of making halfhearted continuous improvement efforts. PlyGem Industries is just one example of many whose attempts at process improvements were reduced to a destructive "Program of the Month" instead of an adventurous journey of continuous improvement involving all employees. Total commitment is what is required. I believe it is possible to utilize simple FlowCycle principles for applying continuous improvement in ongoing total workflow Cycle Time Reduction.

FlowCycle Builds on Cycles of Learning

The purpose of FlowCycle is to imbed cycles of learning into company culture by applying the simple FlowCycle execution methodology of the four phases of MARI – Mobilization, Assessment, Redesign, and Implementation. Every issue facing any organization can be framed in its problem-solving effort around these four phases so that the effort is not haphazard. It is organized and thoughtfully pursued.

Each time MARI is repeated, that becomes a learning cycle. As these cycles of learning accumulate, a culture of learning and change imbeds itself into an organization so that eliminating non-value-added activities in the value stream workflows become a way of life – business as usual. By repeating the MARI process to reduce waste, any company can achieve quantum leaps and World-Class gains in performance through Leaner workflows in the business and manufacturing processes. The number of MARI cycles employees execute drives the rate of change. More cycles equals more change. More employees cycling equal more processes changing.

Cycles of Learning to Transfer Knowledge

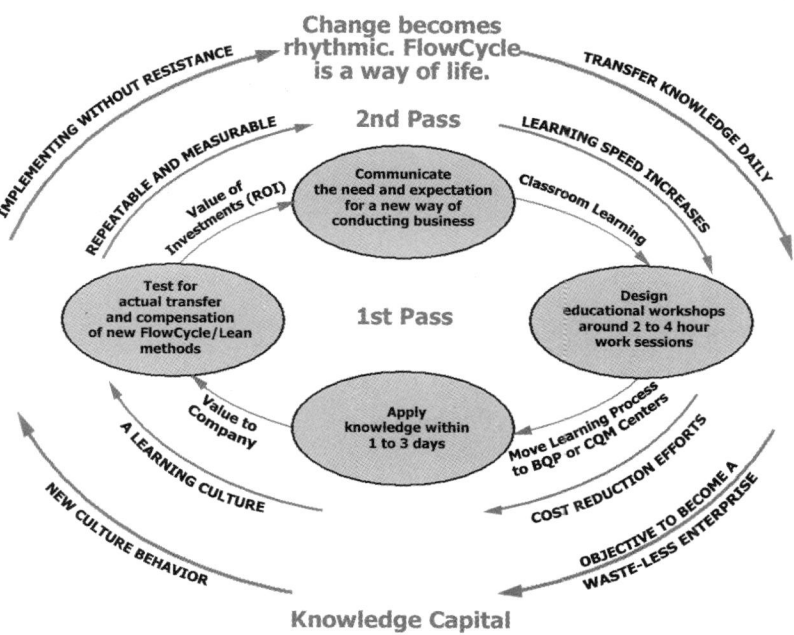

CHAPTER ⚙ TWO

The Three Principles of FlowCycle to Achieving a Waste-Less Enterprise

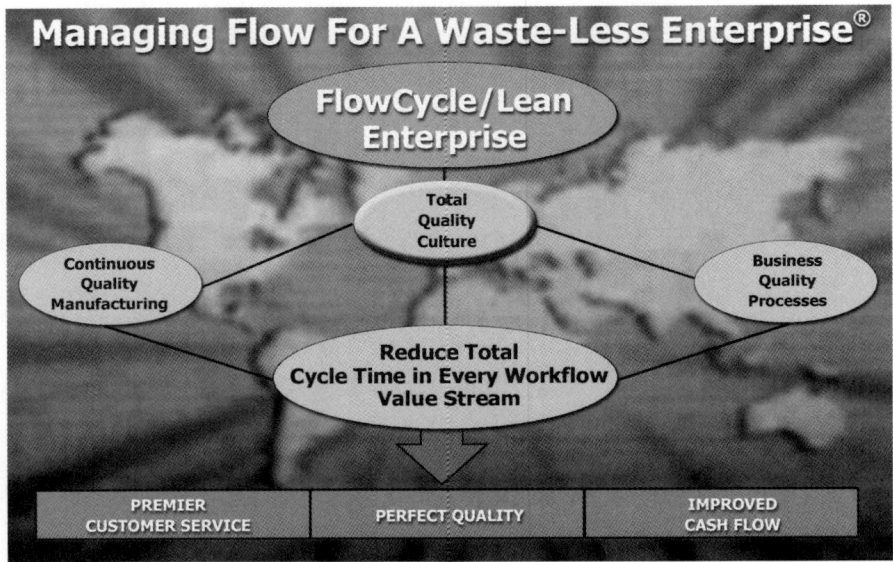

There are three key principles of FlowCycle that create the foundation for a Waste-Less Enterprise continuous improvement journey.

- The first key principle is Total Quality Culture – TQC. TQC focuses on customer requirements and determines what is required to provide instant fulfillment of those requirements. It is a learning culture that puts people first – the best training for leaders, managers, and employees and the best process of learning.

- The second key principle is Business Quality Processes – BQP. BQP focuses on reducing non-value-added time in all business processes. It has the best human resource practices, the best management practices, the best leadership practices, the best IS systems, the best measurement systems and the best order fulfillment processes.

- The third key principle is Continuous Quality Manufacturing – CQM. CQM focuses on all the best concepts of TPS/Lean Manufacturing. It is to have the best in manufacturing practices, focused factories, supply chain management, visual measurements, and quality every time.

The Three Qs are the foundation of true workflow for continuous improvement.

PRINCIPLE ONE – Total Quality Culture (TQC)

*TQC is the foundation for creating **an atmosphere of change** by developing the knowledge of cost of quality within an organization as the key driver for change. This knowledge triggers the need for change in behavior within your company . . . thus converting old practices into new practices and developing people with new skills that **enable them to make change rapidly.***

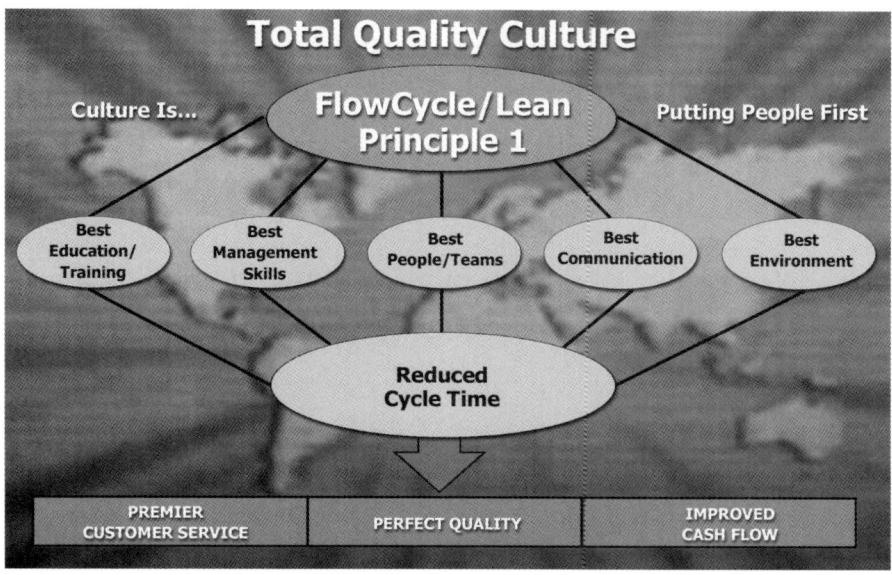

The first "Q" of FlowCycle is Total Quality Culture (TQC). Before describing the new attitudes and behaviors necessary to TQC, we must first understand what is culture and how change and transition work together to improve that culture. To put it simply, culture is the set of values, behaviors, and characteristics of an organization. It is the personality of the company. All improvement activities must be organized around the simple principle of creating Total Quality Culture.

TQC is the foundation for transforming organization culture. TQC starts at the top and must be the overall focus of all performance and reward systems. Continuous improvement programs fail because participants lack commitment to the program. The only way to ensure commitment is to make it an integral part of the company's compensation and reward system.

What is change and how does it happen? Change is transition. Change is individual. It can include a new location, a change of mind-set, a new management procedure, a new team, a new role, or a new policy. Anything that is different than the current state.

How do individuals cope with transition? Transition is the psychological process that individuals go through when coming to terms with change. Change happens externally when an individual makes the transition internally. The following graphic illustrates the change process.

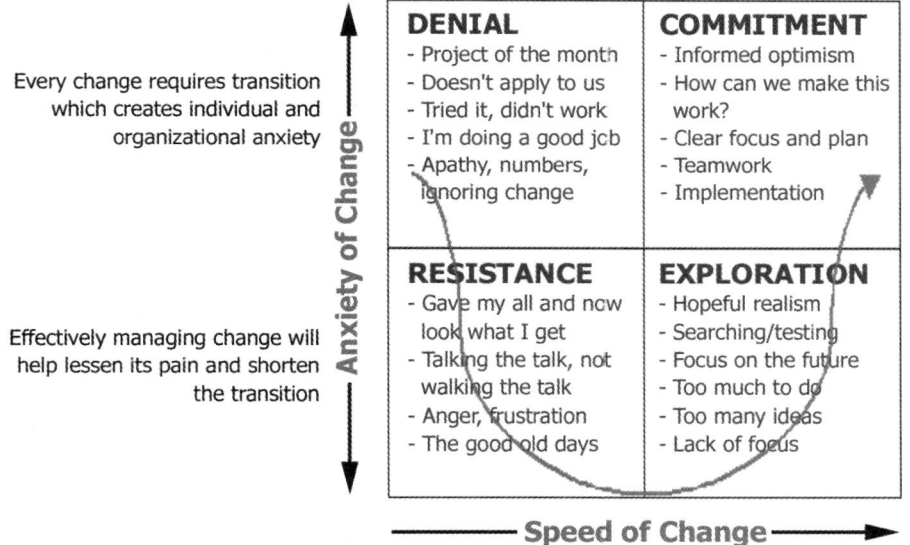

When first faced with change, individuals or organizations must pass through an ending state that brings closure to the old ways. This enables them to move on to the next phase of change where they enter the neutral zone. In the neutral zone, individuals go from denial and resistance to exploration and finally to commitment. This returns the person back to a beginning state, the state where real change can begin.

Change Skills – Not People

TQC does not believe the old myth that "there would be very few problems in the workplace if workers would simply perform the work as they have been directed."

In reality, the overwhelming majority of errors and problems are workflow process problems, not people problems.

The focus must be on managing the workflow processes, not on changing employees. People have a passionate inner drive to be useful and successful, but poor processes frustrate that drive to succeed. Removing barriers and smoothing workflows grants people the freedom to do what they wish to do, which is to succeed. Unfortunately, many employees spend most of their day coping with frustration caused by non-value-added processes. They will resign themselves to the situation only because they feel a lack of power to repair the process problems that annoy them every day.

TQC requires leaders to create a learning environment that identifies gaps between actual and required skill levels and empowers employees to receive training in the required skills immediately upon discovery of a skill gap.

TQC requires teamwork as a non-negotiable component of the continuous improvement journey.

Bad processes are the problem, and teams of empowered, enthusiastic, and skilled people are the solution.

People at all levels must work in rhythm to make a Lean organization work. Everyone must be involved, contribute ideas, and help implement solutions.

Effective human resource development brings to the surface each individual's desire to work, improve, and make a difference. People nearly always want to do the right thing and always want to participate on a winning team. People are not fearful of change when provided with the skills and education to improve their knowledge and taken through the change process in an orderly fashion.

TQC requires leaders who are facilitators of culture change and who encourage and depend upon innovative ideas and opportunities generated by their people. A CEO's adaptability to quick change and his or her ability to evaluate change opportunities are the skills that will carry the organization ahead of others in the twenty-first century.

TQC allows companies to transform traditional management practices into new management practices even before the need arises. Many large companies appear to go on reorganization binges. Something happens and then all hell breaks loose for a while. Employees come to expect these frantic moments and tend to view management with disdain for what they perceive as hysterical reactions. Employees adopt a "this too shall pass" mentality, roll their eyes and say, "Oh, we are just reorganizing again."

TQC principles require constant team building, eliminating the need for reorganization. Management earns trust over time through consistent words and actions.

Waste-Less Enterprise leaders walk the talk. The senior management of a company must set up the culture for success through vision, values, and beliefs that enable Cycle Time Reduction to take hold. This allows constancy of purpose and alignment so that an organization can focus on the customer through its people. Every single customer is at risk every single day, and every potential customer is ready to become a customer every day. The single biggest determining factor in the outcome is the attitude of the one person that customer interfaces with that day.

Hallmarks of a FlowCycle Culture

The following statements reflect the qualities of a company that has achieved a true Total Quality Culture.

- Every person shares in the design of the organization's culture. The culture is created from the behaviors and actions of everyone.
- Culture evolves and each culture has its own unique characteristics. As time goes by, TQC principles will gain power and intensity. Customers will grow stronger.
- Culture is capable of controlling individuals through accountability matching.
- Accountability is a core characteristic of desired behaviors.
- TQC is nonthreatening, penalty-free communicating and participating in a problem-solving culture.

The Cost of Quality

As a company transitions toward Total Quality Culture, numerous paradigm shifts will occur. Total customer satisfaction starts with Total Quality Culture and is the first principle of FlowCycle. Total Quality Culture is the foundation of FlowCycle. The beliefs and the behaviors of people enable organizations to thrive in the marketplace of the twenty-first century. One of the best ways to encourage change is teach the cost of quality to all employees across the entire company. Relating the cost of quality elevates employee understanding that time is money and speed matters. This presentation is a critical part of the Mobilization effort, and in most cases, should be conducted by an external consultant experienced in FlowCycle.

Develop the knowledge of the cost of waste within your organization:

– Use this knowledge to drive the need for change in behavior within your organization

**The Center of Excellence is used for team meetings at all levels.
Each team has an identification area displayed on the wall,
which contains a team picture, project schedule, need helps for review,
and the 6 supporting metrics of FlowCycle.**

The cost of quality is a simple calculation of the cost per minute of operating the business. For example, a company with 100 employees working one eight hour shift per day, 250 days per year, has a grand total of twelve million (12,000,000) minutes. If a company's total costs are $20,000,000, the cost per minute is $1.67. The point when presenting this number is that every minute an employee spends is valuable. Wasted time and wasteful processes damage profitability and endanger customer relationships.

The cost of quality can be improved thirty-five to sixty percent by having everyone focus on improving the processes. Helping people understand the cost of a minute actually improves productivity in ways that actually help your business and manufacturing.

When one starts looking at the improvement of a process such as a new methodology, it is shocking how few metrics one has and how poorly defined they are in determining the return on investment. One needs appropriate and well-defined cost metrics to drive behavioral change and help your employees understand cost trade-offs. For example, every minute converted into a new sale will provide job security and avoid layoffs.

PRINCIPLE TWO – Business Quality Processes (BQP)

BQP is a principle that reduces business process cycle time by the planned, continual elimination of inefficiencies and waste while simultaneously increasing customer-recognized value, internally and externally. This is accomplished through the four-phased FlowCycle methodology, MARI, designed to achieve a quantum leap in business performance and position through **redesigning core processes using the same toolsets as manufacturing.** *FlowCycle principles and methodologies can be applied in the same manner as manufacturing using the same improvement tools, starting with* **Mobilization** *of the organization and* **Assessment** *of the value stream through visual mapping to create* **Redesign,** *achieving faster* **Implementation** *of change.*

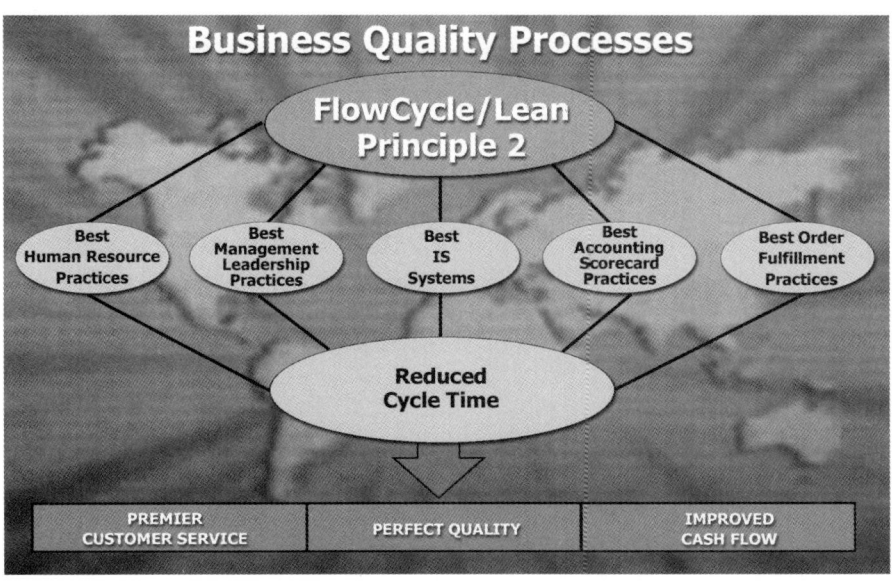

The second "Q" of FlowCycle is Business Quality Processes (BQP). BQP contains some elements of what was once called "Reengineering," although it is a more sound approach focusing on the quality of every workflow process instead of focusing on headcount reduction.

Over time, Reengineering degenerated into a hatchet methodology that radically reduced numbers of people in business processes, leaving the survivors scurrying around to retain customers in spite of the reduced process.

BQP focuses on the flow of each activity to ensure the fastest cycle time when responding to internal or external customer requirements. BQP believes people are the solution, not the problem.

Business processes are service processes such as sales, marketing, scheduling, finance, order entry, engineering design, human resources, payroll processes, product design, IT, etc. A business process consists of a group of logically related tasks using resources to provide defined results that support organizational objectives. As in a manufacturing process, every business process has a workflow, and that workflow has a perceived required time and perceived amount of value-added activities and waste.

The Two Key Business Areas for Change

- Cost accounting must become Lean accounting
to drive new accounting practices.

- Human resources must change all performance plans
to support a Waste-Less Enterprise.

BQP focuses on eliminating non value-added activities in the value stream workflows. Breaking down functional walls is essential in improving business processes and creating new work cells for better and faster customer service. A faster and error-free workflow ensures higher levels of customer satisfaction.

Non-value-added activities divide into two classes:

- Essential business value-added: Non-value-added activities to customers, but may be necessary because of internal or government issues.
- Pure non-value-added: Activities unnecessary to maintain business. These are the areas to eliminate in radically changing and improving core workflows.

Launching a BQP improvement effort requires the support of top management.

There are six times the dollar savings available for recovery in improving business processes versus the dollar savings available for recovery in manufacturing processes.

Teams analyze selected processes and supporting technologies. Business process waste is much more difficult to see than manufacturing waste. Inventory stacked to the ceiling is visible to all, but the costs associated with time wasted in looking for missing files is much harder to capture. FlowCycle makes all waste visible through the simple mapping tool used in MARI.

Using the four phases of MARI, a BQP improvement journey begins by mobilizing the team and defining customer requirements. Teams develop present-state assessments and detailed process flow maps to identify the inputs and outputs of all activities – and whether they are business value-added or pure non-value-added. Pure non-value-added activities become the initial focus for identifying improvement opportunities. This entire process is detailed in Part II.

BQP is the continuous reduction of cost and cycle time by reducing non-value-added time from core business processes:

- **Sales and Marketing**   - **Procurement**

- **Information Systems**   - **Human Resource Practices**

- **Business Administration** - **Financial Measures**

A Waste-Less Enterpriser does not start with Lean Manufacturing – there are many workflow processes that occur before there is any expenditure of labor. There is 6 times the savings in non-manfacturing value processes.

PRINCIPLE THREE – Continuous Quality Manufacturing (CQM)

*CQM is an **ongoing examination/improvement** effort, which ultimately requires integration of all elements of the value stream of the manufacturing processes to achieving the lowest-cost, defect-free product, delivered on time and completed in the **shortest possible cycle time.***

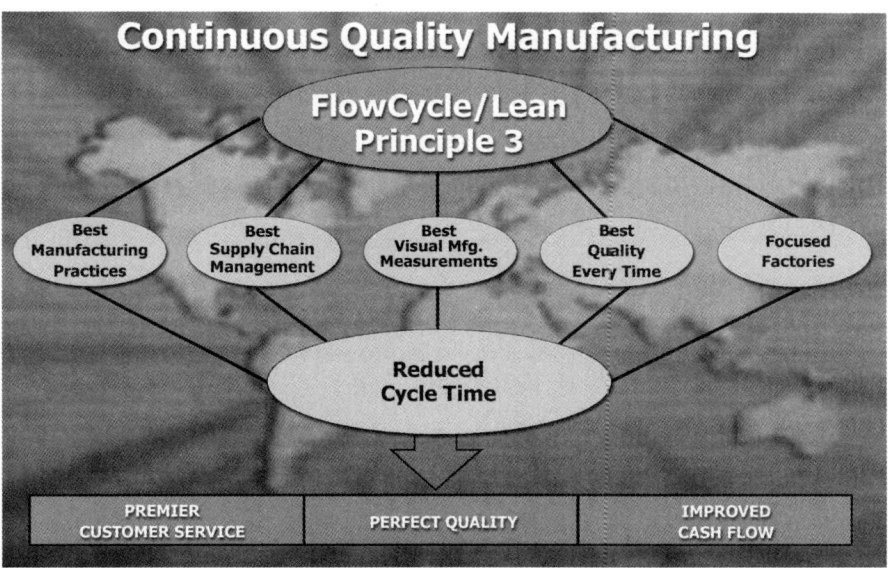

The third "Q" of FlowCycle is Continuous Quality Manufacturing (CQM). CQM utilizes Toyota Production System, JIT, Kaizen, and IBM's Continuous Flow Manufacturing methods, tools, and techniques to produce the right product, on time, every time.

CQM first focuses on bringing discipline and stability to the manufacturing process through a Kanban/Pull System. In a FlowCycle implementation, Kanban (pronounced KON-BON, not CAN-BAN) means "what to build, when to build, and how much to build." Literally, Kanban means a "card" in Japanese. Cards or other visual devices trigger operators with signals so they know what to build, when to build and how much to build.

The reason for implementing a Kanban/Pull System at the outset is to bring discipline and stability to the process. Kanbans reveal true bottlenecks faster than any simulation program. When improving processes, Kanbans usually begin simply, and as sophistication increases, so do the Kanban calculations. Manufacturing employees transitioning to a Kanban/Pull System environment require support and training.

Creating the "Visual Factory" is crucial for successful implementation and maintenance of manufacturing excellence. Each manufacturing sector will maintain CQM boards for visual measurement, value stream maps, and daily and weekly schedules. Continuous Quality Manufacturing means building the right product on time, every time.

These basic workflow process improvement tools are based on value stream maps and cycle-time analysis. There are many tools that might be used. These tools are generally relevant to Cycle Time Reduction and can be even more broadly applicable to culture, business processes, and manufacturing processes.

- Kanban/Pull Systems

- Set-Up Reduction

- Defect Prevention

- Operation Time Improvement

- Process Time Improvement-Cells-Flow Lines

- Total Productive Maintenance

- 5S's

CQM is the continuous reduction of cost and cycle time from all manufacturing processes by eliminating:

- Unplanned downtime = part shortages, machine break-downs, absenteeism, transportation delays, etc.

Management walks the process from shipping area back through the process to understand daily constraints of unplanned downtime that is prohibiting you from meeting customer requirements. The coach/supervisor or any team member can explain the reasons behind these unplanned constraints at their CQM center with factual data.

The three principles are the firm foundation for implementing and sustaining continuous improvement. Part II of this book describes the MARI execution methodology as a detailed playbook for managing the journey to continually reduce cycle times throughout all workflow processes in the value stream.

FlowCycle, the MARI methodology, brings visual management and waste elimination to every process and builds a communication system that lets employees share and celebrate in the results of their efforts.

PART TWO

FLOWCYCLE
MARI
METHODOLOGY

→ MARI – Organizing and Implementing Change

FlowCycle™ Triangle

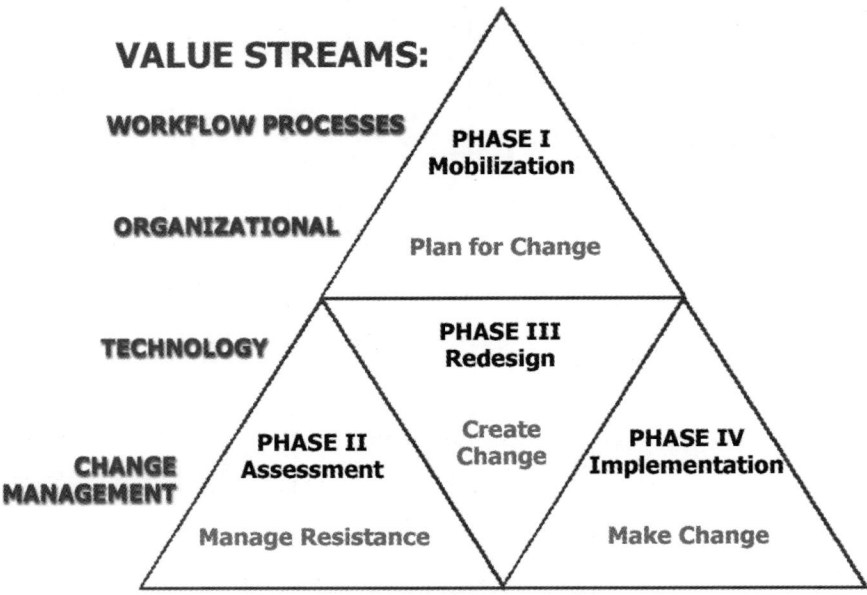

VALUE STREAMS:

WORKFLOW PROCESSES

ORGANIZATIONAL

TECHNOLOGY

CHANGE MANAGEMENT

PHASE I
Mobilization

Plan for Change

PHASE III
Redesign

PHASE II
Assessment

Create Change

PHASE IV
Implementation

Manage Resistance

Make Change

Your company begins by mobilizing resources and ends by implementing workflow process redesigns based on customer requirements.

FlowCycle, the MARI execution methodology, is designed as a framework for organizing and implementing Cycle Time Reduction to achieve a lean enterprise. The goal of FlowCycle is to imbed into an organization's culture continuous improvement as a daily operational process. The MARI triangle illustrates the process an organization will go through that is repeatable, measurable, and repetitive.

Mobilization is the first phase, at the top of the triangle, because it has to do with people, and people are a company's most important asset. Mobilization focuses on the involvement of people, setting the foundation, selecting the critical core processes, and launching a successful process of change by communicating the company vision to all employees. It is the leaders who must plan change. It is the leaders who will manage employees' resistance to change. It is employees who design and create the ideas for change, and it is employees who will actually implement change. That is why the focus of FlowCycle is people and why the first phase is all about mobilizing people to change behaviors.

Mobilization is the top of the triangle because assessing, redesigning, and implementing change is impossible without first preparing the employees for change.

Leaders must demonstrate the need for change, the benefits of change, what will change, and how change will impact each individual. Mobilization lays the foundation and readies the people involved for new ways of thinking and working. Mobilization aligns employee activities and tasks to the company vision established by leadership. Mobilized employees, now ready to embark on a change journey, must clearly and visually see the workflow issues in their processes before they can begin creating solutions.

Assessment is the left corner of the triangle and the second phase. Assessment is one of the most important phases of FlowCycle. The first step of the Assessment phase begins with analyzing both internal and external customer requirements. The second step is completing a current-state analysis of your workflows, whether they are business or manufacturing processes. These current activities are then broken down into "business necessity" and "waste." By using visual mapping tools, employee teams clearly see the waste in their processes and challenge old assumptions. This causes resistance, and leaders must channel resistance energy into redesign energy to create new ideas for the future state. Improvement efforts can fail in this phase as people try to justify and revert to the old ways. Leaders manage this resistance by continually demonstrating their commitment to improvement and reminding everyone of the urgent need for change.

Redesign of those processes becomes the third and next logical phase. The Redesign phase is upside-down, because during this phase if you are not driving cost out and reducing cycle time, you should stop at this point because there will be no improvement to customer performance. Setting the Redesign goals and priorities by utilizing the information from the mapping process is important as the first step of Redesign. This ensures when you create the new design or improve upon the old design that you take into consideration what occurred in the total Assessment phase. Prioritize those improvements that are "quick hits" or "low-hanging fruit" to ensure commitment to change. Enthusiasm builds for improvement as teams envision life after change and begin to see the possibilities of creating something new and dynamic. Redesign is not a time for timid tweaking.

The final phase is Implementation – change made real. The implementation triangle points back up to Mobilization because once change is sustained, it is time to repeat the cycle of MARI. Implementation is not easy. It can be done haphazardly, which would revert the organization back to old ways.

Successfully implementing change requires the development of a detailed Implementation action plan. Pilot and test the new value stream workflow processes and measure the results against the goals. Then roll it out for continuous improvement with the involvement of all employees. Implementation makes the effort of Mobilization, Assessment, and Redesign worthwhile. Improved cash flow, energized employees, and satisfied customers make implementation exciting.

This methodology was designed for repeatable, measurable, and sustainable performance to avoid having to invest in a new program year after year. In fact, all of the education that I have designed is based on building blocks. Building blocks ensure that you attain results within a dramatically short time period. How long does a MARI cycle take? Of course it depends on the amount of the waste, but from a macro perspective, most companies I have coached achieve dramatic, meaningful results in less than four months. Overall workflow Cycle Time Reductions of 50 to 75 percent are common.

The FlowCycle MARI methodology is conviction-driven. It involves a thorough transfer of knowledge, and in the words of Don Shula, makes you, "audible-ready, consistent in your approach, and cost-of-quality driven."

Beginning a Continuous Improvement Journey: The Needs Assessment

Every successful journey begins with planning and preparation. Management must determine, from a strategic standpoint, where the business currently stands and set goals for the improvement journey. Often management knows there are problems but does not have the toolset to attack the problems in an organized way. The lack of an effective problem-solving framework manifests itself in many ways. Feeling the urgency to "do something," management may take one or more of the following actions:

- Throw money at the constraints.
- Withhold money from the constraints.
- Issue across-the-board cost-reduction edicts.
- Issue downsizing mandates.
- Hire people to throw at the constraints.
- Purchase training modules, or send people to training.
- Hire expensive consultants.
- Spend a fortune on software that contains 60,000 features and options.
- Purchase real estate and equipment.
- Move production outside the United States.

While some or all of the above may be prudent actions, the first thing to do is to assess the needs of the situation in detail. Only then can a leader make a wise, informed decision about how to proceed.

The Needs Assessment is the initial element of the FlowCycle methodology. It is performed to determine the current state of affairs and to set goals for the future state. The Needs Assessment is the first element of Mobilization because it gives management a clear statement of the issues. Management gains an understanding of the path and cost of a successful journey. Most important, they gain the knowledge that all is not lost. There is a solution. It can be done and it will not take all that long. Management is beginning to be mobilized. A Needs Assessment has three major components: Culture Assessment, Operations Assessment, and a Journey Plan. The Journey Plan includes a proven project schedule and this becomes the playbook.

The Culture Assessment

A successful continuous improvement journey requires every player on the team to pull together in accomplishing a common vision of "what good looks like." The Culture Assessment identifies the gaps in understanding between the people who work in various functions of the business. Identifying misunderstandings, communication issues, management style issues, culture and team behaviors allows company management to target training and improvement strategies to the most critical issues. The Culture Assessment always produces fascinating results that make it obvious to management that culture issues exist, which require immediate action.

The Culture Survey

In a FlowCycle Needs Assessment, all employees of the facility complete a simple, one-page Culture Survey.

The survey typically reveals that while management may communicate company goals, employees do not understand the goals.

Managers nearly always feel they have communicated their vision and goals repeatedly. People listen, but many times never hear and understand the message. This lack of understanding has its roots in the actions of previous managers, and may have nothing to do with the current management team. Regardless, people only hear and understand when they believe the information presented has a significant impact on them as an individual.

> Empowered employees listen, hear, and understand.
> Powerless employees do none of the above.

At a FlowCycle implementation at a process factory in the Midwest, the Culture Survey results were very revealing. This factory was located in Green Camp, Ohio, where the jobs provided were extremely important to many families in the area. The facility had been losing substantial amounts of money for many years. It had twice changed owners before being purchased by a large conglomerate in 1988. After eleven years of losses, the parent corporation had come to the conclusion that making a profit from this facility might not be possible and set a timeline to consider closure. Fortunately for the hardworking people at the plant, there was a business unit manager who, like Moses, prayed to his masters for patience with the people. This manager was given an ultimatum: "Fix it, or we're shutting the plant down in six months.

The Needs Assessment results as a whole, and the Culture Survey results in particular, surprised management. Management had always blamed the problems at the facility on the union. Actually, the problem was extreme amounts of waste in the workflow processes. All the waste had resulted in constant, intractable problems that predictably caused finger-pointing between management and the union.

> People are rarely, if ever, the problem. Wasteful processes
> are the problem. People only become a problem when
> bad processes prevent them from succeeding.

Understanding the communication gaps between various levels of the organization is the most important area to concentrate on when embarking on a journey of change. This is why the first section of the survey starts with communication.

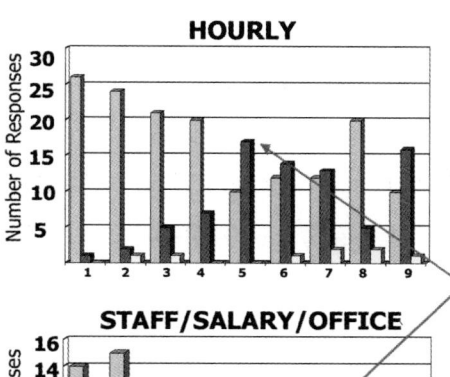

(A) COMMUNICATION

1. Does your company communicate its goals?
2. Do you think attaining these goals is attainable?
3. Does management communicate organizational changes?
4. Does management communicate why organizational changes are important?
5. Do you see any business measurements?
6. Do you understand the measurements?
7. Do you have the right information to do your job?
8. Does your supervisor or manager have a positive reaction when you ask for additional information?
9. Is conflict considered good or bad?

Since the change journey will build around team empowerment, gaining an understanding of the current teamwork environment is the second critical area covered by the Survey. Teams will learn what skills are required for actually implementing change, not merely talking about it.

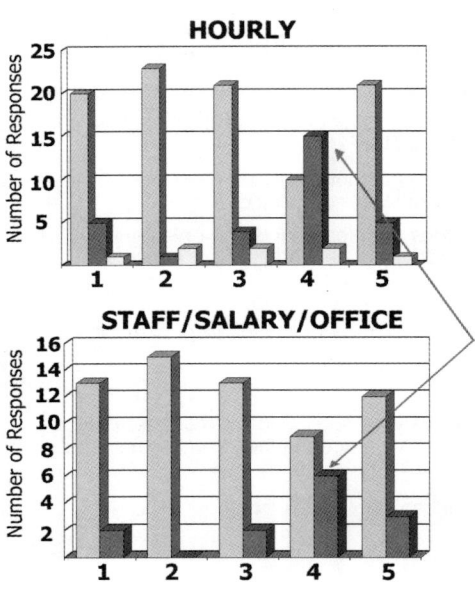

(B) TEAMWORK

1. Have you worked on a team here?
2. Do you enjoy working as part of a team?
3. Has teamwork improved the business?
4. Have you had team training?
5. Is teamwork valued at this company?

The Survey's third area involves understanding attitudes toward empowerment. In some cases, team empowerment has gone too far by allowing teams to set 100 percent of their own goals. In other cases, employee empowerment has been given lip service only, not really allowing people to be involved at all. After understanding the gaps of this area, management in Mobilization will define all the team empowerment levels. They will define empowerment into two simple areas – Can-Do's and Need-Helps.

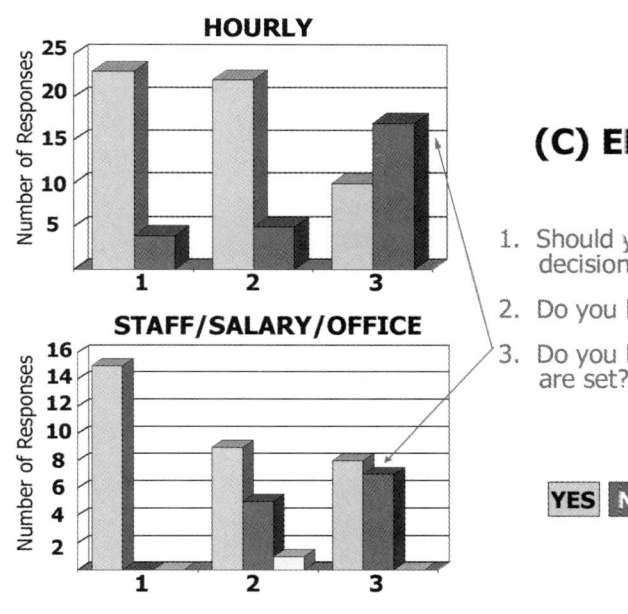

(C) EMPOWERMENT

1. Should you be able to make more decisions about how to do your job?
2. Do you know how priorities are set?
3. Do you have input on how priorities are set?

Understanding the attitudes of an organization toward its leaders and the style of those leaders is the fourth area of this Survey. This knowledge is crucial to building empowered teams that communicate successfully.

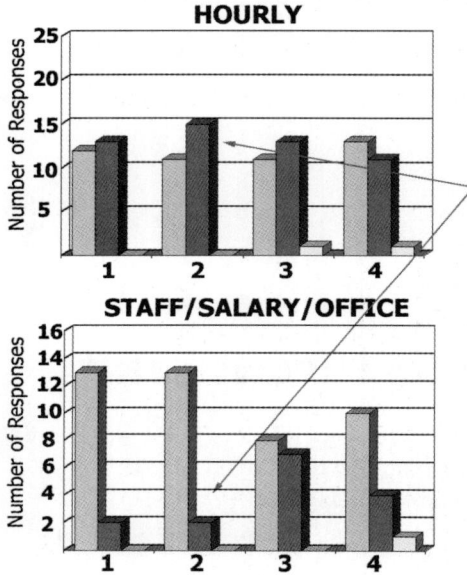

(D) LEADERSHIP

1. Do you consider your direct supervisor or manager to be a good leader?

2. Does your supervisor inspire you to be a great performer?

3. Does your direct supervisor or manager give you opportunities to develop your job skills?

4. Does your direct supervisor or manager clearly communicate expectations to you?

Understanding employee attitudes toward feedback and recognition is the Survey's final area. Feedback and recognition work both ways. People want to be told in a respectful way when something is wrong. They also need to be told when something is right. It always comes back to the first part of the survey – communication. People respond more quickly when they hear their name and receive timely feedback. Being a team member and hearing their name may be the only incentives to offer employees in recognition for their effort in this journey. After all is said and done, knowing someone's name has lasting meaning.

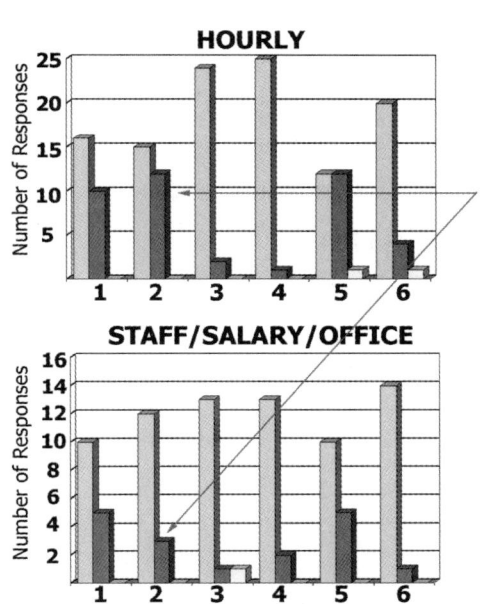

(E) FEEDBACK & RECOGNITION

1. Are you rewarded or recognized for contributing to company goals?

2. Do you receive feedback on how you are performing?

3. Is feedback constructive?

4. Is feedback important to accomplishing your job?

5. Are you rewarded for doing a good job?

6. Do you give your manager or supervisor feedback?

The graphs above show the questions asked and illustrate the gaps in understanding between salary/office and hourly employees.

At the factory in the Midwest, the Culture Survey yielded an understanding of many positive changes that had been previously suggested in the company and gave an understanding of what percentage of the people would support change. This led to an understanding of the root causes of poor schedule adherence that had hindered achieving shorter cycle times in previous improvement efforts.

Culture Surveys nearly always reveal that people are afraid to take risks. Why? Because risk requires trust, and that gets to the heart of leadership styles.

Hourly and business personnel often believe that management lacks credibility to make fundamental process changes because of "programs of the month" that have come and gone like a Texas tornado.

They believe their supervisors lack the team leadership skills required to actually change anything. Why? Because many times management and supervisors have come up through the ranks with no formal leadership training. This does not mean these people are bad leaders. It does mean they may need new skills, not on how to manage, but on how to coach in order to lead in new ways.

Management responses to the Survey will often reveal a lack of project management and execution skills needed to execute change rapidly. Leadership styles vary, and management may migrate toward a continuous improvement process, but management often feels overwhelmed by the prospect of implementing a change initiative. Why? Because they know they still have to get the normal job done with its daily firefighting, so they fear failure. FlowCycle overcomes this fear by providing the plan and infrastructure that ensures a successful and continuous change journey.

Properly assessing culture issues in an organization is critical to designing and implementing a successful journey plan.

The good news at the factory in the Midwest is that profits are now beginning to flow, like water from the rock struck by Moses. While the union and management may not yet see completely eye-to-eye, the small town has an employed population of management and employees that respect each other.

Operations Assessment

The second major component of a Needs Assessment involves a process walk-through and the creation of high-level maps of the core processes and products for improvement.

In a manufacturing environment, engineers experienced in the application of Lean principles can visualize what a Lean factory will look like based on a walk-through of the existing plant. Similarly, the core business processes can be understood and mapped by simply asking the people involved in the process to explain what they do. Not surprisingly, they will explain where all the constraints exist, and most of the time they have ideas as to how to remove them.

The high-level value stream maps must clearly highlight the barriers to smooth flow through the value stream. The key principle of effective maps is that they must work from meeting the customer requirements back through to the beginning of the process.

One of the common problems associated in meeting customer requirements is shipment linearity. A lack of consistent production and shipping is clearly illustrated by a graph of daily or hourly production. The graph will be shaped like a hockey stick with a sharp upward curve at the end of a period. One goal of Lean is level shipping and production rates with product pulled through the factory. This can be seen in a high-level map of the processes associated to the non-value-added time in the value stream of the workflow. The maps enable a preliminary design of a generic Kanban/Pull System in the manufacturing or business process to instill discipline and consistency in a process.

Kanban/Pull Systems were developed in the Toyota Production System for managing correct inventory levels given the characteristics of the entire manufacturing process. Kanban is a Japanese word for "visual card." The system was actually developed by a German grocery chain in California. The vice president of manufacturing of Toyota was on vacation when he observed a set of cards in a holder behind the last case of food on shelves. When he saw this, he called it a "Kanban." The visual record had information for the store clerk to know where to pull replenishment stock from the warehouse. Pull systems control production by pulling product through the factory or business processes at the rate of customer demand, no more or no less than what is needed.

The value stream map allows experienced eyes to see where cells and flow lines can be recommended and designed for future state improvements. It allows for management of the Takt rate, the time required to meet customer demand. Takt literally means the "rhythm of the heart." Your heart does not ship all its blood at the end of a week – neither should a manufacturing or business process. The goal is a steady, predictable flow of requirements pulled through the processes at the pulse rate of customer demand.

For conducting a Needs Assessment on a business service company, a typical high-level process map that identifies constraints is shown below.

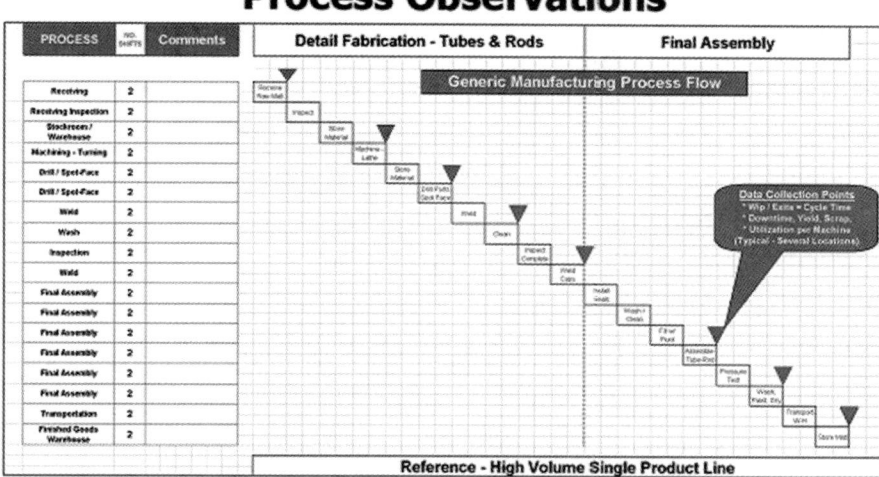

The key to successfully observing opportunities for process improvements is to bring in people experienced in Lean implementations to help create the vision of "what good looks like." As Henry Ford observed in 1926, "The easiest of all wastes, and the hardest to correct, is the waste of time, because waste of time does not litter the floor like wasted materials."

The Journey Plan

The Journey Plan includes examining the expected impact, the cost of hiring external coaching, and the payback or return on investment from the journey and a preliminary project schedule. A well-designed and implemented continuous improvement journey will increase productivity and reduce rework, scrap, overtime, inventory, floor space, labor, and overhead costs. Estimates of these improvements are then compared to the overall estimated costs of the journey over the next year to determine a return on investment.

There are many intangible benefits of the FlowCycle journey because of the transformation of the culture of the company from one of distrust into one of dynamic learning and teamwork. The tangible financial payback should easily exceed three to one with a realistic expectation of a ten-to-one return in the first year.

Management rarely hesitates to purchase new equipment based on an ROI of three to five years, but mysteriously they hesitate when told they can achieve a 3-to-1 ROI on the investment from a FlowCycle improvement methodology in the first year by getting every employee focused on improving workflows and reducing costs. They hesitate only because they do not really know what good manufacturing or business workflows look like. They also do not understand how all the improvements can take place virtually "overnight" in a few short months. They seriously doubt their ability to deliver, even if they have the knowledge. Management is often paralyzed by the fear of failure. All this fear is unnecessary, because the FlowCycle methodology removes the guesswork and uncertainty by laying out a simple playbook and teaching players the plays.

> This is the FlowCycle difference. FlowCycle methodologies
> supply the tools, training, and project plan to break the
> journey down into a series of simple, manageable phases.
> FlowCycle prevents analysis paralysis by supplying an
> execution playbook and then programming the sequence
> of the plays. The result is continuous action on the
> part of everyone toward the goal of Cycle Time Reduction,
> elimination of waste, and implementing workflow
> improvements. It is this relentless continuous action
> by everyone that produces rapid results.

Implementing a successful continuous improvement Journey requires an irrevocable personal commitment from upper management. Making the commitment is difficult, but for the courageous and wise, the benefits are enormous.

We've seen an over 30 percent reduction in the labor required to produce our product. With Flow Management's MARI techniques, a very rapidly growing company would not have to hire additional workers. When I first met these guys, I would never have believed this possible. With the MARI approach, you get benefits at the beginning of a process – not the end of a process. I expect additional benefits the over next several years as we take their methodology and continue to implement it. They have given us a vehicle with which to implement continuous improvement.

Brett Olson,
President & CEO Black & Decker, Brazil

Are you ready to write your own success story? Let's get started by learning the four secrets of success – MARI – Mobilization, Assessment, Redesign, Implementation.

CHAPTER THREE

FlowCycle Phase I
Mobilization

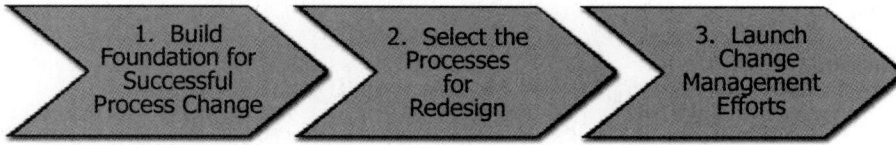

| 1. Build Foundation for Successful Process Change | 2. Select the Processes for Redesign | 3. Launch Change Management Efforts |

Creating change starts with Mobilization of the whole organization. Mobilization prepares the people for new ways of thinking, working, and behaving. Mobilization involves everyone because everyone must become involved to create continuous cycles of improvements. Change is not solely the job of management. It is everyone's job.

This section discusses:

- What is change?

- What elements cause change to occur through the three steps of Mobilization?

- How do people and organizations react to change?
- What positive changes occur as a result of launching the process and getting people outside their comfort zones?
- Why do people resist change and how do we overcome resistance?
- What is the secret of successful, sustainable change management?

Mobilization is the critical first phase for successfully implementing redesigned workflows and improvements in the customer value stream. Often, this phase is neglected altogether or given inadequate attention. I will begin by making some points that many leaders may be tired of hearing. Leaders say we cannot afford to have all our people on teams or this-that-and-the-other team is not necessary. However, the plain fact remains (consistent with basic respect principles) that with constancy of purpose, one can build a company with a common set of beliefs, which will result in more customers. If the people in the organization are aligned and motivated with this belief, the competition is beat. Many organizations attempt to change using other improvement processes but fail because they bypass this phase.

Stressing the importance of mobilizing all employees to create change challenges many of the basic skills that have made many managers successful. Successful Mobilization calls for a great deal of hard work and new attitudes among leaders to break old habits and embrace empowerment and inclusion.

Cycle Time Reduction is much more than moving
a few machines around in a factory. It involves
transforming people's roles and behaviors,
which then transform the organization.

In Deming's famous 14 Points, he emphasizes, "substitute leadership" (he mentions this in two of the 14 Points). Good leadership steps to the plate and makes the hardest decisions that provide hope. Why? Because 97 percent of a company's problems are likely to be waste in operational systems and 3 percent are likely to be organizational problems. The point is to make the concept of leadership come alive throughout the organization by ridding it of those problems. The issue is not new technology, hardware, and software. The issues are how to change business practices, operational procedures, management skills, and organizational behaviors.

Mobilization builds upon the culture gap information learned in the Needs Assessment to build a foundation for successful process change. I have the honor of speaking at the roundtable of many management functions. I hear managers say we practice "servant leadership," but actions speak louder than words. I see leadership practicing command and control behaviors. High-level value stream maps help management identify the value streams to select the process for Redesign and improvement. The maps will make visible the type of management behavior that is driving the organizational performance. This is the time to launch change management efforts, creating a sense of urgency throughout the organization. Some leaders may feel as though nothing I have talked about sounds new. People are the foundation and they are the most valuable asset. Only accounting practices keep companies from putting them on the balance sheet.

Step 1: Build Foundation for Successful Process Change

Effective change management is a journey, not a one-time motivational meeting. Each day presents new circumstances, challenges, and opportunities, and leaders have the responsibility, opportunity, and privilege to continuously guide others to embrace change and to set the correct behaviors. A Total Quality Culture focused on total customer satisfaction must be able to adjust and change instantly. This is a key to becoming a market leader and widening that lead.

Understanding Change

Let's look at the change loop again, beginning with the end in mind – that is, "What does good look like? What does World Class look like? How can your company be World Class?

Change Loop

ENDING	NEUTRAL ZONE	BEGINNING
CLOSING STATE		CHANGE OCCURS

Characteristics of the Transition State:
- **Low stability of individuals and organizations**
- **Perceived high levels of inconsistency in the workplace**
- **High emotional stress - expect a sense of confusion, ambiguity**
- **High, often undirected, energy - sense of loss**
- **Emphasis on control of people vs. processes**
- **High value placed on past behavior patterns**
- **Increased conflict - deterioration of trust**

The purpose of the culture survey from the Needs Assessment includes giving people the opportunity to vent frustrations and identifying cultural barriers. The object of mobilization is to get people into the neutral zone of transition to start adapting to new ways of thinking and working.

> Often during the transition to the neutral zone, emotional stress, confusion and a sense of loss appear. Many times undirected tension will surface with the closing out of attitudes. Leaders must learn that it is the processes causing problems, not the people.

Mobilization seeks to eliminate the high value placed on past behavior patterns that increased conflict and eroded trust. The closing phase is to eliminate instability in individuals and organizations.

Change management helps reduce stress to those who are changing. This enables people to assimilate into a new culture within a shorter period of time. The greatest opportunities for change are created out of situations involving pain or pleasure.

Human Change Curve

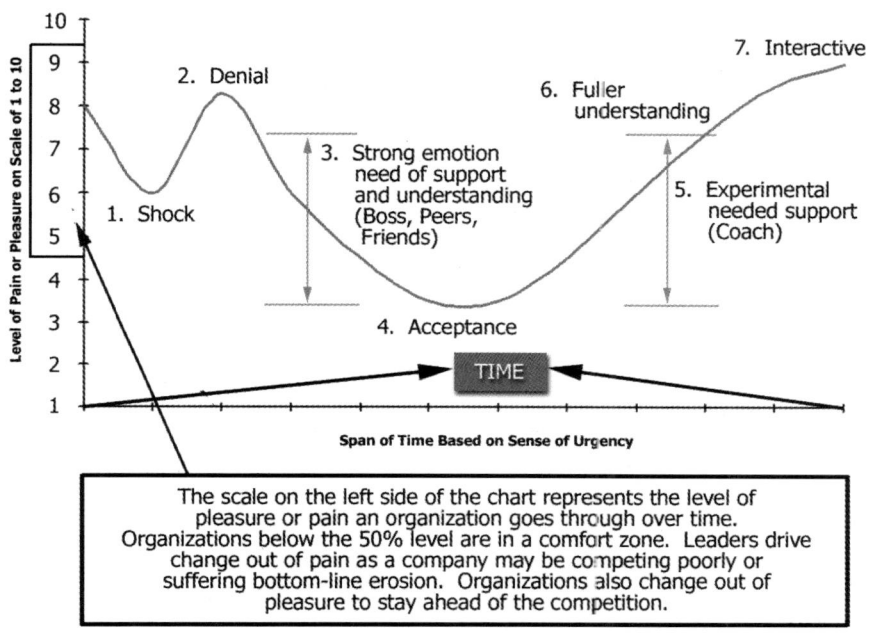

The scale on the left side of the chart represents the level of pleasure or pain an organization goes through over time. Organizations below the 50% level are in a comfort zone. Leaders drive change out of pain as a company may be competing poorly or suffering bottom-line erosion. Organizations also change out of pleasure to stay ahead of the competition.

The anxiety of change (left side of the graph) is determined by the level of pain or pleasure in the organization. The speed of change (bottom of the graph) is determined by what drives change based on pleasure or pain. The chart shows what happens through the phases of change: denial, resistance, exploration and commitment to change. Organizations travel through the change process faster when in pain or pleasure. Everyone goes through the same change process on a personal level in life as well.

The Mobilization phase focuses on change management, not because it is more important than establishing the new workflow process, but because it is the precursor to establishing new improved workflow processes that become sustainable. Often, planning for change is uncomfortable to the entire organization. Change means reestablishing objectives and setting new goals.

A key component of changing behavior is changing everyone's performance plan to include commitment to Cycle Time Reduction as the key driver.

When planning for change, it is important that the journey include all of the company's processes. This involves developing a team organizational structure that integrates all levels of personnel with management personnel.

Steering Committee Team Level

The Steering Committee is the foundation of change management because they must become a solid foundation for team sustainability. Organization structure for this change method should be derived from the current functions. First, one should work on the value stream processes, and then reorganize to match the new value stream workflows in redesign. Membership includes the highest-level manager at a division, plant, or company and his or her direct reports. This committee meets weekly and is led by the leader of division, plant, or company. A FlowCycle/Lean Site Champion prepares and facilitates the weekly meeting with a set agenda.

The Steering Committee will establish a team structure as recommended below.

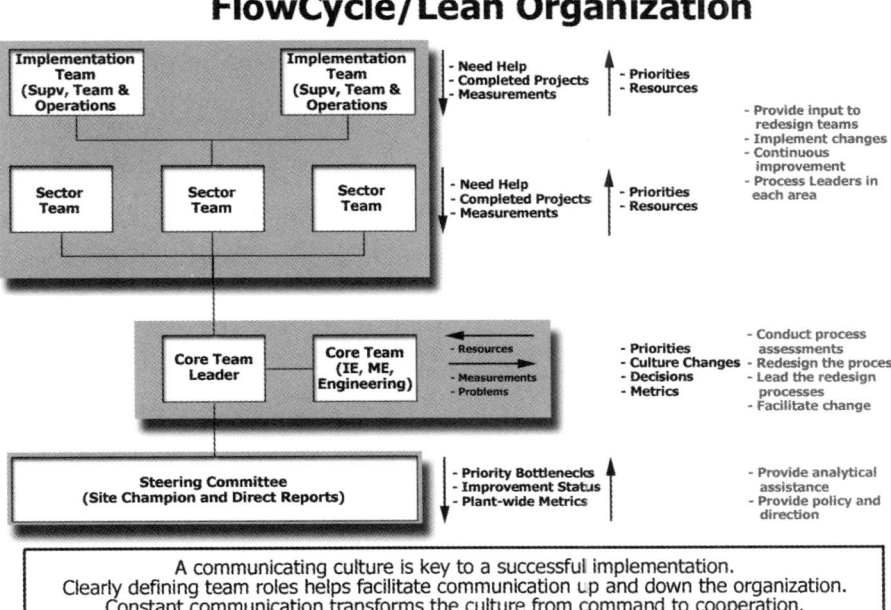

The organization chart shows senior leadership on the bottom of the chart because they are the supporting, solid foundation. The "iceberg architecture" melts down. They must be solid, like concrete, in their commitment to the journey. Leadership supports and sets guidelines, but the employees are empowered to find and make improvements based on guidelines of leadership. By placing leadership at the bottom of the organizational chart, leaders are sending a clear signal that the old autocratic command and control climate is out and the open, communicating culture is in.

Management vs. Leadership

Are "leaders" and "managers" synonymous? Not necessarily. Many of us have worked for managers who made the numbers and got things done, but we wouldn't work for them again if we had to. Those people were good managers but poor leaders.

True leaders are individuals who motivate people to get things done and then give others credit for any success.

An effective technique for training leaders is to have each leader list his or her top three strengths and top three weaknesses on 3 x 5 cards and then share that information with the team. The simple exercise of disclosing strengths and weaknesses to peers is very powerful. This exercise is important to align management level leadership and staff toward the common vision.

Too many organizational leaders surround themselves with people that are reflections of them. When this happens, organizations get stronger in the strong areas and weaker in the weak areas. Excellent leaders instinctively know they must trust others by gathering peers that strengthen their weaknesses. Leaders that build a staff around their weaknesses and understand that they cannot do it all by themselves obtain results much quicker than leaders that build a staff only around their strengths.

Functions of the Steering Committee

The Steering Committee has four major functions:

- Develop and deploy a communications strategy for the direction and vision of the organization.
- Direct the core team's overall Cycle Time Reduction priorities, develop strategic direction, and assure a focused effort to resolve conflict and remove obstacles to achieve the vision.
- Review progress weekly and provide constant visible support to the Journey.
- Develop FlowCycle organizational reporting, roles, how teams are set up, whether they are a manufacturing or a business service company.
- Define how people on the teams operate and what authority the leader has to make necessary wide-ranging changes by appropriate understanding of Can Do's or Need Helps.
- Hold people accountable to the vision and help them maintain a positive attitude for achieving sustainable workflow improvements.

The Vision Supported by Mission Task Statements – Core Values – Core Strategies

In a FlowCycle implementation, the Steering Committee meets for a three- to five-day off-site team building experience. This bonding experience generates commitment to the vision by clearly defining Mission Task Statements. Leaders must personally demonstrate the vision, passion, and direction of the organization with clear communication of the goals and expectations. It is important for the leaders to manage with empathy while retaining the image of strength and authority.

Often, performance plans do not support the leader's vision and so they drive managers in the wrong direction, which creates conflict. The solution is to align the leader's vision with managers' Mission Task Statements and performance plans.

Colin Powell says,

Military is political and I would prescribe to the corporate leadership that they have all the same purpose: to define the visions of an organization, to figure out the missions you need to accomplish that vision, and then to organize people in the most effective way to accomplish those missions and achieve the vision of the organization.

To be successful, there can only be one vision. The Titanic is a great example. The captain set the course to meet customer requirements of a smooth ride, but he failed because of the inability of his staff to follow his vision.

The following story cannot be historically documented, but read through it and imagine its possibility. The Titanic did not sink simply because it hit an iceberg. The Titanic sank because of failure to execute the Mission Task set by the captain. The vision was one of grandeur – to meet the expectations of the customers by providing the smoothest ride on the sea in the atmosphere of a ballroom. The mission was to navigate through a particular area through the night as people celebrated by dining and dancing.

Picture the value stream – the Titanic launches. It is off to sea. During the night, at the change of shifts, the co-captain gets caught up in the activities and the grandeur of the Titanic at sea and leaves his duties at the helm.

Suddenly he realizes the ship has veered off course. He failed to pay attention to the navigation necessary to execute the vision. He attempts to correct the error by increasing the ship's speed and slightly changing course to make up for lost time. Not knowing what constraints lay ahead, the Titanic hit an iceberg.

What happened with the Titanic was the co-captain's hidden agenda of engaging in the fun and games onboard ship caused him to stray from the Mission Task of staying on the course set by the captain. The captain failed to assign an accountability partner to his co-captain to ensure adherence to the vision and Mission Task.

In a continuous improvement journey, attempting
to shortcut the process can have disastrous results
for current and future journeys.

The senior leader must express one clear, concise vision with bullet points for objectives. Consider this example of a vision statement to achieve Lean.

Worldwide Vision of Cycle Time Reduction to Achieve Lean

Provide impeccable customer service through World-Class leadership in sales, design, supply chain management, fabrication, distribution, manufacturing, and support operations.

- *Provide our customers with the highest-quality product or service on time and complete, at the best price.*
- *Continuously improve performance by reducing cycle time in all workflow processes, activities, and tasks.*
- *Create a preferred workplace for energetic, skilled, motivated employees committed to excellence.*
- *Provide profitable growth of our business.*

Once the vision is developed, the question is how to achieve this vision. The vision is pursued through Mission Task Statements, which define tasks for each one of the functional areas reporting to a senior leader. Each functional area has its employees complete Mission Task Statements covering their roles. Share the vision by having every individual participate in creating his or her own personal Mission Task.

Mission Task Statements – Not Sub-Vision Statements

So how are Mission Task Statements created? Think of the word "task." What is the task required to accomplish that vision? Mission Task Statements must be clear and precise about the tasks that are needed to accomplish the vision. They are a critical early step in providing clear leadership to achieve speed throughout the value stream.

For example, a management level Mission Task could be to train all employees. Another Mission Task could be to establish monitors to understand how the body of the organization and processes are working by building tools like a Center of Excellence, Continuous Quality Manufacturing Centers, or Business Quality Process Centers.

Vision for Success	Core Values
Our vision is to satisfy our clients by coaching them through a successful implementation and leave them with knowledge to succeed on their own. We bring people and proven methodologies together to help end users perform successful change. - Keeping our Methodologies simple, - Practicing Lean/Flow internally, - Empowered employees who perform like owners, - Impeccable Customer Service	- Honesty - Openness - effective communication - Trust - Respect of others
Mission Tasks	**Core Strategies**
- Provide direct reports with "Internal Projects" to further develop and expand the services, products, and core competencies of FMA. - Establish a solid infrastructure to support active client engagements, including proactive account management and effective internal/external communication (Internal action items, Project Status Reports, etc.) - Develop Performance Plans for direct reports to guide their personal and professional development.	- Improve requisite technical capabilities through active participation in appropriate training and reading. - Assist in the development and management of a comprehensive Financial and Operations Model for FMA products and professional services. - Further develop our personal knowledge of FMA policies and procedures in preparation for an expanded leadership role.

Each member of a functional or departmental part of an organization should develop his or her own Mission Task Statement in the format displayed above. One task should always be to reduce the non-value-added time and identify the purest value-added path.

It takes clear and precise communication and a dedication to "walking the talk." Why? It is not a leader's job to be the expert at processes within a company. Leaders set goals and commit to those goals and do not change course with the latest fad. For a true leader, walking the talk means staying on course and knowing when someone is trying to bluff about taking action.

Great leaders know the value of a vision. Some business gurus have argued that the fastest, easiest way to change an organization is to reduce its size. False!

A clear vision with leadership commitment and an organized approach to getting everyone involved is truly the fastest and easiest way to create change.

Core Values

Core values are the handful of fundamental, uncompromising principles upon which the people of the business operate. Senior leadership identifies and documents the core values to provide all employees with clear expectations of appropriate behavior. Examples of constructive core values include honesty, open communications, trust, respect for others, professionalism, and a positive attitude. Each individual must be totally committed to these basic values. Misalignment between company and individual core values is a formula for failure.

Core Strategies

Core strategies are the primary professional/personal initiatives to be undertaken to support the vision by each employee during the next 6 to 12 months. They are listed in the "Core Strategies" quadrant of the chart. The individual determines the basic components of a personal contribution plan in regard to the objectives of the organization. In other words, "What can I do during the next several months to increase my long-term value to the business?" The answer is to improve one's self to meet the requirements established by the vision, but it is important to be specific.

Once the vision is established, leaders must continuously monitor and motivate managers to vigorously pursue the vision and produce the needed behavioral changes. They should spend time working actively with managers and coaches, monitoring, supporting and motivating to remove the barriers to change. They should spend a considerable amount of time on external relationships with the customers. Leaders must communicate up and down through the organization about the need for change that will remove the barriers to achieving speed through the value stream and allow the company to compete at World-Class levels.

Core Team

A specialty team leading the improvement journey is the called the Core Team. The Core Team's membership includes the site champion, key managers, and technical resources that meet both on a daily and weekly basis. The Core Team's role is to establish the training schedule, set up the training room, and deploy all the tools needed for each team to be successful.

The Core Team is also charged with developing high-level value stream maps used for determining which processes or products should be selected for

Redesign at the company level and support key improvements. The Core Team develops the program management tools to assist in the success of all teams posted in the Center of Excellence.

The leader of Core Team is also the Site Champion. What kind of person effectively leads a FlowCycle/Lean journey?

A Site Champion is someone respected enough to enforce compliance of all teams to follow the Journey Plan. Of course, almost everyone talks about so-called "special teams" – teams of co-located people reporting to a Core Team leader with broad responsibilities and authority. This often is a reaction to optimize for cross-functional efficiency but has lots of political, resource allocation, motivational, and other problems that prevent projects from getting done rapidly. Often the Site Champion must avoid giving into inside company politics and stay focused on the success of the journey.

The Site Champion must have the power to demand that people be held accountable through personal performance plans. The Site Champion must always show his or her passion and be absolutely committed to the journey.

Sector Teams

Sector Teams are formed from functional areas that include the departmental managers. A sector is usually a department or function within an organization. Sector teams meet weekly with the leaders of the Implementation Teams they monitor. The Sector Team is charged with determining how to design and deploy the process improvements at a plant level. They carefully monitor the performance trends recorded on the Continuous Quality Manufacturing and Business Quality Process centers, provide direction to the Core Team in process improvements, and support Implementation Team improvement projects as needed.

Changing an organization's culture is a serious process, but it need not be difficult or take long. The key is establishing and defining team roles throughout the organization. These seldom follow a linear path through an organization.

Working relationships are highly complex and seldom can you create enough change by simply having cross-functional teams.

Each leader needs to identify those individuals who are positive and those who are negative. Leaders must then build upon the positive stakeholders' positions to encourage others to open their minds to change and break through the barriers erected by the uncooperative.

The logic for setting up a journey in this manner is to gain everyone's support for the change. This frees the Core Team to focus on the major constraints of the value streams. Each departmental team legitimizes its contribution by finding and implementing workflow improvements that do not negatively affect its suppliers or customers. While the Core Team focuses on major constraints, Implementation Team members contribute to the overall success of the journey by mapping and then eliminating the non-value-added activities and tasks within their manufacturing or business process.

Ultimately, each sector leader is responsible for overall Implementation Team results and for communicating improvement activities to the Core Team and Steering Committee.

Implementation Teams

Direct supervisors lead the Implementation Teams composed of the supervisor and those employees who work in the details of the process, whether a business or manufacturing process.

Implementation Teams meet weekly. The teams review the previous week's measures and plan actions to meet the current week's customer requirements. Implementation Teams receive training in the form of two- to four-hour educational modules of the four phases of MARI. This process of education provides multiple cycles of learning. This is why teams document their efforts using Continuous Quality Manufacturing (CQM) or Business Quality Process (BQP) centers. Both centers provide critical information for continuous improvement and bring visibility of team activities to leadership.

CQM or BQP CENTERS

FRONT
- "Can Do's"
- "Need Helps"
- Daily Graphs
- 6 Key Charts

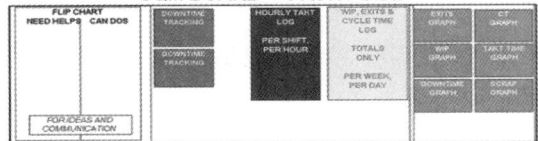

BACK
- Process Issues
- Activity Worksheets
- Process Map
- Transportation Diagram

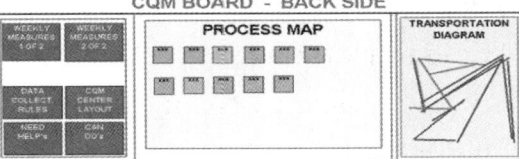

The front side of the CQM or BQP center, on the right-hand side, records the six key measures documented on a daily basis.

The center of the board records the weekly, daily, and hourly requirements needed to meet customer schedules.

The left-hand side holds a pad of paper used for establishing and documenting Can Do's and Need Helps. Can Do's are improvements the team can implement independently as defined by the rule set forth by the Steering Committee. Need Helps are improvements that require assistance from the Sector Team, Core Team, or Steering Committee. Often, dollar amounts are set for Can Do's. For instance, if the improvement will cost more than $500, then it falls into the Need Help category.

On the center back side of the board, teams will create a detailed map that ties to the overall value stream map of the order fulfillment process developed by the Core Team.

- The back left-hand side of the board, records brainstorming ideas.
- The back right-hand side holds a transportation diagram of the process. The transportation diagram shows who supplies their process and who the customers are. It visualizes the waste of transportation by showing the travel distance of product as it moves through the process.

Team Ground Rules

The Steering Committee establishes team ground rules for success. The ground rules include:

- Establishing team-meeting agendas to drive change at each level. Agenda formats are the same at each level and are designed to drive change based on reportable metrics.

Rules as to length of a discussion of a particular.

- Defining penalties for noncompliance with rules. Penalties can be financial, such as $5/minute for late arrival for management level meetings. At Implementation Team levels, the penalty may be to sing the alphabet, clean the meeting room, etc.

- Documenting guidelines for Can Do's and Need Helps for all teams.
- Rules on everyone being on time. No one leaves early.
  No one interrupts.
- All team agenda are formatted to focus on measures first, then key drivers of cycle time improvement.

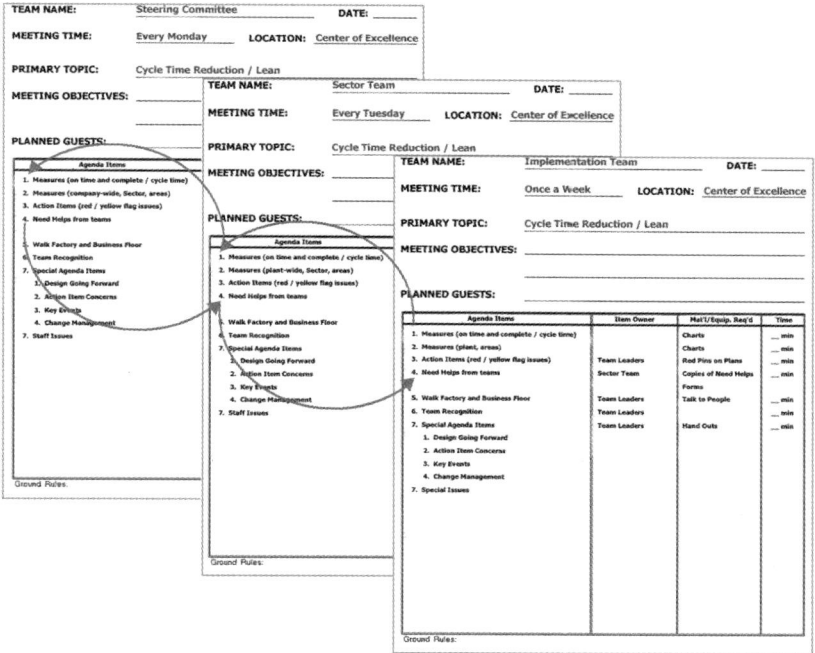

- Identifying key support and process improvement initiatives in developing program management tools within the sector to monitor how teams are achieving Cycle Time Reduction to become Lean.
- Validating its team membership with key members at all levels in managerial and technical resources needed.
- Assigning accountability partners.

It is important at the end of each meeting to briefly critique the meeting. As the meeting is being critiqued, teams need to look with openness and respect toward each other, to establish what was good and what could be improved.

Accountability Partners and the Four Styles of Behavior

An important element of Mobilization's Step 1 is to assign accountability partners. Training facilitators assign accountability partners based on a behavioral style analysis done to ensure that the individuals on the team bring out the best in each other.

In a FlowCycle implementation, coaches use the behavior styles material developed by Dr. Tony Alessandra. His relationship strategies are very simple and fun. A quad chart of four behavioral styles is shown in the figure below.

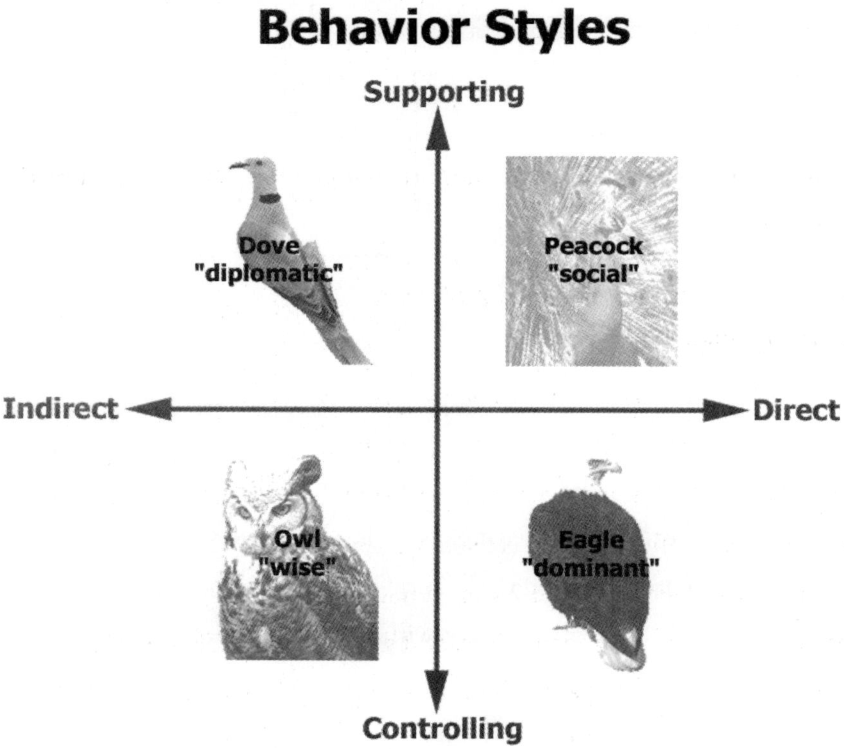

Behavior Styles

Supporting

Dove
"diplomatic"

Peacock
"social"

Indirect ←——————————————→ **Direct**

Owl
"wise"

Eagle
"dominant"

Controlling

The characteristics of the Dove as a Relater:

- Diplomatic, supporting, indirect, unassertive, warm, and reliable
- Seeks security, makes decisions slowly
- Avoids risk and is people-oriented
- "Notice how well-liked I am"
- Actively listens and loves to build trust

Doves must learn to occasionally say no. They are best paired with a mid to strong Eagle or a mid to direct Peacock.

The characteristics of the Owl as a Thinker:

- Indirect and controlling
- Analytical, persistent, and a problem-solver
- Aloof, picky, and critical
- Needs to be right, asks a lot of questions
- Likes deadlines, details, and is well-organized
- "Notice my efficiency"

Owls must learn to show concern. They are best paired with a mid to direct peacock or mid to weak Eagle

The characteristics of the Eagle as a Director:

- Controlling and direct
- Highly productive, bottom-line results
- Likes challenges, but cool and independent
- Subordinate, but also impatient and very tough on other members
- Impatient, inflexible, and is usually a workaholic
- "I want it done right and I want it now."

Eagles need to listen, pace themselves, and show concern. They are best paired with a midrange Dove, a midrange Owl or a midrange Peacock.

The characteristics of the Peacock as a Socializer:

- Direct and supporting
- Fast-paced, spontaneous, and fun
- Energetic, generous, on time
- Enthusiastic, persuasive, influential in the process
- Entertainers, risk-takers
- "Don't confuse me with the facts."

Peacocks need to control their emotions and objections. They are best paired with a strong Eagle or a Dove, who can be direct.

The best behavior style for the Site Champion is a Peacock or a borderline Eagle. The purpose of assigning accountability partners is to draw the best out of each team member by purposefully pairing them up with other members on the team with a different style. The goal is to achieve balance on teams and hold members accountable to completing tasks and changing behavior. People's behavior style and how they communicate impacts a process flow positively or negatively.

The purpose of accountability partners is to speed implementation of process improvements by eliminating personality clashes, cliques, and hidden agendas and to ensure completion of tasks on time.

The person facilitating the training assigns accountability partners based on behavior style and should avoid assigning friends together. You should never pair Owls or Doves as accountability partners.

Organizing for Success

Organizing for success is the final element of Mobilization: Step One "Build Foundation for Successful Process Change." The quad chart below summarizes the central ideas of Step One.

Understanding the four key areas of the quad chart in the following order visualizes the vital concepts for the team. The top left quad emphasizes time and speed.

The concept of time is the most important philosophy of all.

Time is speed and speed is the competitive edge.

Leaders must make Cycle Time Reduction business as usual.

Make Cycle Time Reduction Business as Usual	**Organize for Success**
- Combine all aspects of improving the business with running the business. - Site Champion co-leads change vs. executing project tasks. - Site Champion team members establish peer relationships with your team.	- Deploy and develop all resources consistent with FlowCycle - M.A.R.I. Methodology. - Align organizational structure around goals to achieve results. - Align ownership, accountability and responsibility at all levels. - Insist on early wins for visible commitment.
Steering Committee Must Lead The Charge	**Demand Results**
- Align vision, goals and performance. - Champion change - change of this magnitude cannot be delegated. - Develop a sound decision making process and use it. - Communicate and share all known information with the enterprise as soon as practical - this practice yields TRUST! - Fully embrace a learning and problem solving culture.	- Create a sense of urgency - "Results" - Identify, link and monitor performance - Align and link rewards to results for teams - Prioritize and control projects in process - Demonstrate a bias for teams that take action.

Leaders must lead by example. Senior management cannot simply delegate this process to people who have no power. It is best if the Cycle Time Reduction journey starts with the CEO. The CEO can create one vision for all employees.

Then everyone at all levels aligns his or her Mission Task Statements with the overarching vision for maximum speed to enhance shareholder value.

Organizing for success means investing in team members by providing learning opportunities early and often throughout the journey. Unpleasant organizational changes are often necessary, and leaders must demand accountability at all levels by changing personal performance reward systems.

Leaders must demand results and be prepared to make tough decisions for the good of the company if results are not satisfactory. Leaders must insist on measured bottom-line results driven by real measured wins.

Step 2: Select the Processes for Redesign

The second step of Mobilization is "Select the Processes for Redesign." In Mobilization, processes must be prioritized based on each workflow process' overall cycle time in the value stream. The following checklist helps select a process or product family that will yield the targeted results.

Use the following checklist to help your team select Value Stream Workflow Processes that will yield the targeted results.

☐ The process/product family is related to key business issues.

☐ The process/product family directly impacts on external customers.

☐ The process/product family has visibility in the company.

☐ The process/product family consumes many resources.

☐ Management agrees that the process/product family needs to be improved.

☐ Management will cooperate with and fully support the cycle time reduction efforts.

☐ The process/product family is not in transition or scheduled to be overhauled.

☐ The process/product family has clearly defined supplier and customer starting and ending activities.

The diagram below shows the relationship from the time the customer orders to the time the customer receives a product on time and complete. In this process, there is a customer need, a sales order process, a manufacturing process, distribution, invoicing and collecting, and finally, delivery of the product to the customer.

TOP LEVEL VALUE STREAM OF ORDER FULFILLMENT

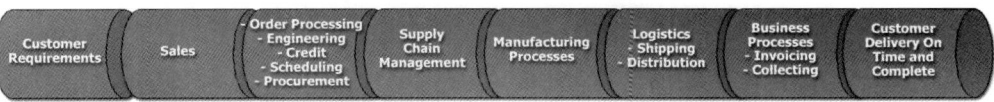

Examples of interrelationships under the high-level value stream include the supply chain management processes along with after-sale services and support processes. These are all the actions required to bring a product or service through its workflows, door-to-door, from raw materials to the customer's facility, on time and complete.

The map below illustrates how high-level processes begin to interrelate and drill down into detail. Cycle times of each activity and the process as a whole must be calculated so that areas for improvement become apparent. This cycle time data keeps team members focused on areas where the most benefits can be derived. As always, improvements start with activities closest to the customer.

ORDER FULFILLMENT WORKFLOW PROCESSES

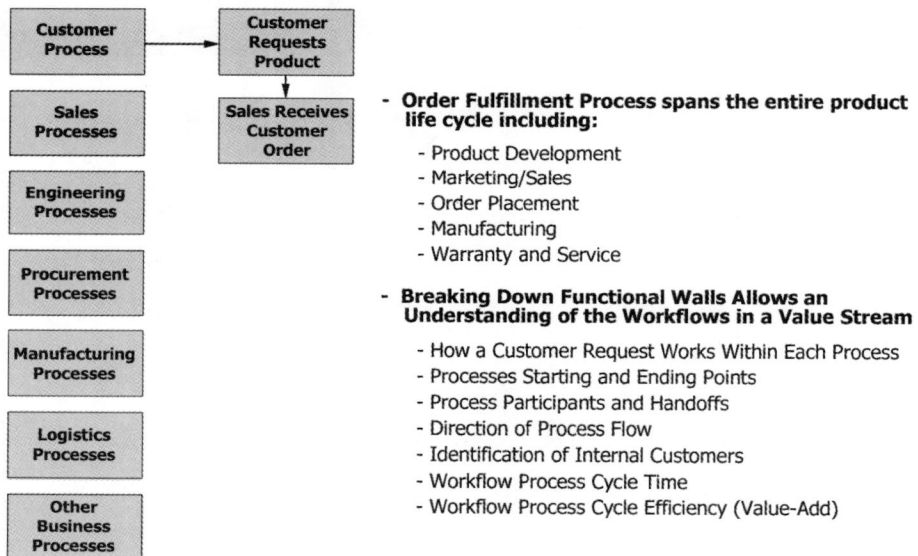

- Order Fulfillment Process spans the entire product life cycle including:
  - Product Development
  - Marketing/Sales
  - Order Placement
  - Manufacturing
  - Warranty and Service

- Breaking Down Functional Walls Allows an Understanding of the Workflows in a Value Stream
  - How a Customer Request Works Within Each Process
  - Processes Starting and Ending Points
  - Process Participants and Handoffs
  - Direction of Process Flow
  - Identification of Internal Customers
  - Workflow Process Cycle Time
  - Workflow Process Cycle Efficiency (Value-Add)

Examples of constraints include: shortages, product development, high scrap and rework, long setup time, unclear work instructions, and no documented design of manufacturing ability or assembly. It could be supplier certification, unreliable suppliers, long distance of suppliers, or processes within the factory. It could be lack of standardization, a high number of suppliers, no ownership or accountability, line measures, large or wrong batch sizes creating excess inventory or organizational issues. It is important to map out each process and calculate cycle time to identify root causes.

Order Fulfillment Process Steps

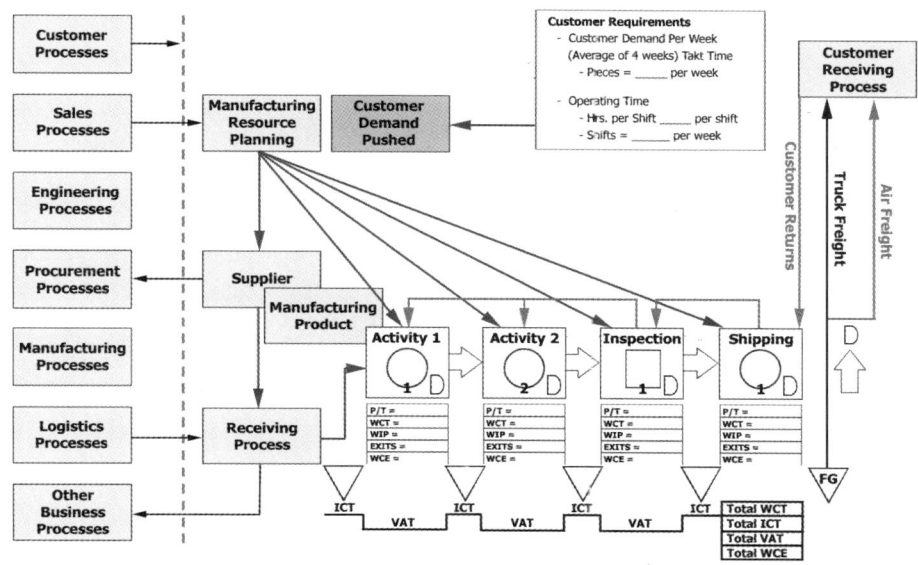

Maps assist in designing Kanban/Pull Systems for inventory reduction, meeting customer demands, and building the right product at the right time. Maps visualize work in process and point-of-use storage locations. Maps facilitate the design of Lean costing systems for labor, inventory management, accounting, and cash flow improvements.

The high-level map is required to design the necessary team structure because it visualizes all of the processes, and how all internal processes interrelate, and it shows the touch points for suppliers and customers.

In a FlowCycle journey, a value stream map lists the components of the organization down the left-hand side and the activities in sequence from left to right, flowing from the top left down to the bottom right. The overall cycle time is tracked for each key sector, then each activity is tied to a team's detailed process flow map.

Information from the high-level map provides the data to create Pareto charts by area visualizing the longest cycle time constraints. This is important for two reasons:

- In a manufacturing environment, it helps design and implement the first level improvement through a Kanban/Pull System that ensures adequate inventory at each sector to meet the needs of the downstream customers' Takt rates.
- The Pareto diagram identifies sectors with the longest cycle time. The longer cycle time sectors have the most opportunity for improvement, therefore these areas become the target areas the Core Team investigates for constraints and bottlenecks.

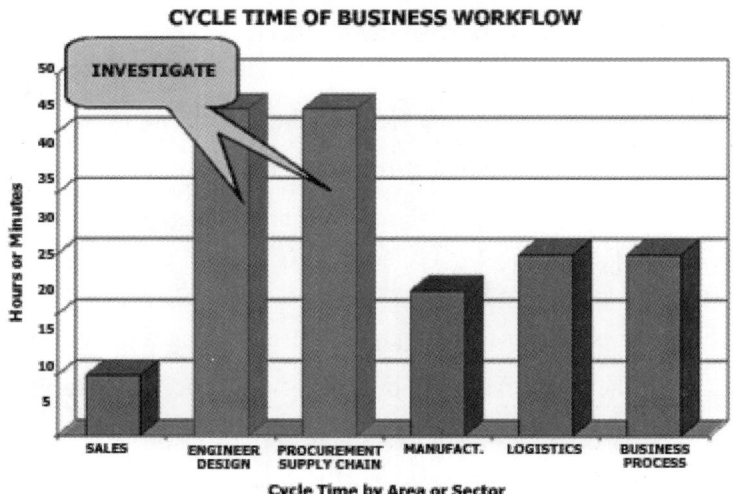

CYCLE TIME OF BUSINESS WORKFLOW

Cycle Time by Area or Sector

The FlowCycle methodologies allow your team to determine where to start in the total value stream or redesigning functional operations into separate value streams by processes or by product families. In this case it looks like the engineering and procurement supply chain through the BQP centers need further analysis.

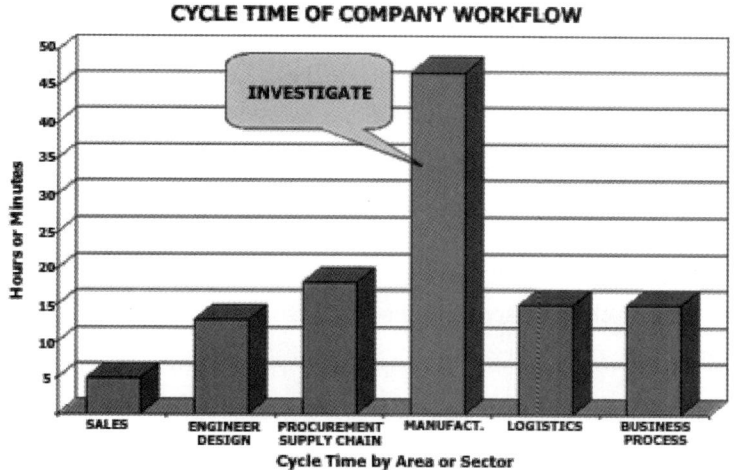

CYCLE TIME OF COMPANY WORKFLOW

Cycle Time by Area or Sector

In this example, the manufacturing sector has the longest cycle time. The CQM centers in manufacturing with high level value stream maps need further analysis to understand if this is driven by product families or process families.

Step 3: Launch Change Management Efforts

In Step 3, "Launch Change Management Efforts," the infrastructure for creating the visual tools that manage flow is deployed throughout the facility. These visual tools are absolutely essential to sustaining the journey. Without continuous, enthusiastic maintenance of these tools, the journey will begin to flounder, culture gains will be jeopardized, and management credibility can plunge.

Accountability is visibility, and documentation
is a key factor in successfully achieving the vision.

The Center of Excellence room (COE) is the intensive care unit for display and monitoring of all team activities. Team training and meetings take place in the COE. The COE brings visibility to all sectors of the company. The Center of Excellence room contains a series of boards that visualize the following:

- Journey Schedule, the master project plan for launching and continuing the journey.
- The high-level value stream map that visualizes the total order fulfillment workflow of a company, plant, or business process.
- Part of the Center of Excellence is sectioned off for the Implementation Teams. Each team updates its board using the following list:
  - Team Journey Schedules.
  - Six key measures for tracking improvements.
  - Lists of barriers to resolve.
  - "Need Helps" requirements and data.
- Team results and recognitions to build pride among teams.

CENTER OF EXCELLENCE
(COE Room)

PURPOSE
- **Contents and Layout to Monitor the Pace of Change**
  - **Legend by Color Coded to Create Visibility and Accountability**

Visibility

Creates

Accountability

The Center of Excellence room brings visibility to the journey. The schedule color-codes each one of the functional areas to identify who is responsible for which specific components and tasks of the journey.

Information Systems

Financial

R & D

Engineering

Manufacturing

Human Resources

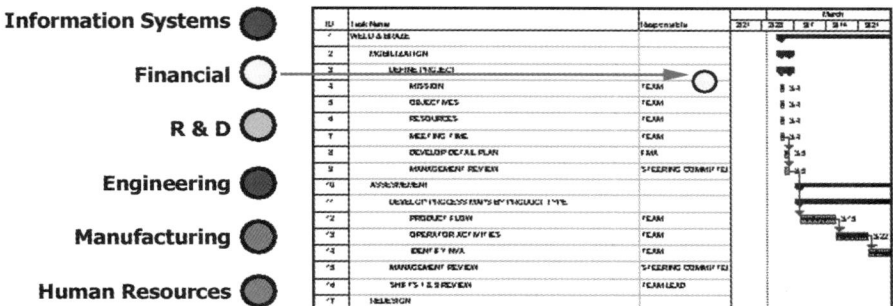

Each corrective action can have more than one functional owner. Each function can quickly identify its action by color pin in the left hand column.

The Red Flag Process – For Guaranteeing Results

The diagram illustrates a journey schedule, a corrective action legend, an inter-functional action list, the agreed time line of functional actions, the action's required date, the action to be completed, and finally, a very powerful process called the Red Flag Process. The red flag highlights tasks that are not being completed on time and identifies the persons causing delays in the journey schedule. If a functional owner agrees to a particular date for achieving a task and fails to complete that task, a red flag is hung next to that activity in the Center of Excellence next to that team's project plan.

The red flag stands out like what it is, a red flag, and sector leaders know where and who the problems are. The Steering Committee now can manage by exception and resolve barriers as soon as they appear.

Visibility ensures accountability and
accountability ensures action.

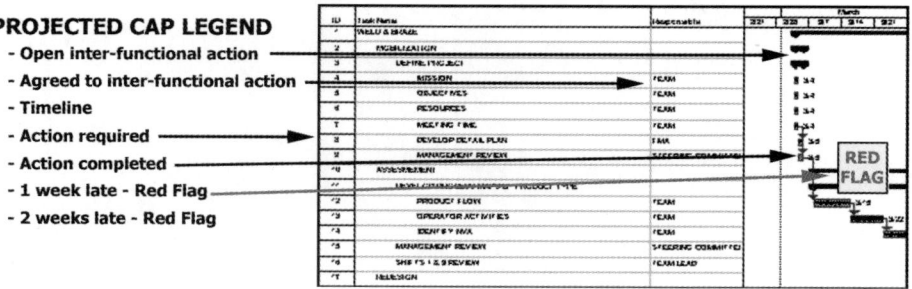

PROJECTED CAP LEGEND
- Open inter-functional action
- Agreed to inter-functional action
- Timeline
- Action required
- Action completed
- 1 week late - Red Flag
- 2 weeks late - Red Flag

As teams begin making process improvements, the bar of performance will continuously be raised higher and higher. Productivity will skyrocket and productive capacity will be unlocked with no additional resources, people, or machines being committed to the process.

Sector leaders will need to motivate to the hilt, create a sense of urgency, recruit good people by creating a supportive work environment, pass out "psychological paychecks," listen, encourage, and recognize jobs well done. Praise and recognition for small and large improvements are critical success factors. All these efforts reduce job stress and create a sense of rhythm and harmony.

Mobilization Conclusion

Without successfully mobilizing the employees, any continuous improvement effort will quickly lose momentum and run out of steam. It will become another "program of the month," and management credibility will suffer another blow. Leaders must manage themselves and play multiple roles.

Leaders must become part motivational speaker,
part Cycle Time Reduction evangelist, part therapist,
and part industrial engineer and, most important,
leaders must be consistent and believe in the journey.

Lou Holtz, currently the head football coach of the University of South Carolina Gamecocks, is my favorite example of someone who has a proven plan that works over and over, no matter where he coaches. Holtz has now taken a record six different schools to postseason bowl games.

One of Holtz's strangest techniques is to understate how good his team is playing. After South Carolina beat ranked Mississippi State on Saturday, September 24, 2000, to improve to 4 and 0, Lou Holtz said, "We just aren't a good football team right now. We've got to get better fundamentally. We have got to break down the barriers and improve our blocking for our running game. We can't stop anyone from running. We just struggle right now."

Translation? Watch out! Lou Holtz has never inherited a winning team, but Holtz has a game plan and it works.

As a college head coach, whether at William and Mary, North Carolina State, Arkansas, Minnesota, Notre Dame, or South Carolina, he has always taken his team to a bowl game in the second year. This was Lou Holtz's second season at South Carolina and the Gamecocks won their bowl game. After a winless season in 1999, Holtz and South Carolina broke a 21-game losing streak.

What did Lou Holtz do to get this team off to such a great start in his second year? He motivates his players to believe in the team and themselves. Holtz says, "I tell you right now, they should make people vote before they have a drink. What I mean by this is that everybody has to look at the turmoil within his or her life and overcome adversity."

Organizations and cultures throughout the world have proven that they will not only follow leaders with a clear vision, but they will follow with passion. Your companywide launch can be just as exciting as the South Carolina fans with the foundation of a committed management team.

Holtz sees adversity as a repeatable process, which he can use to achieve dramatic results.

As Lou Holtz overcame the adversity of his mother dying and his wife discovering for the second time that she had breast cancer, he still adhered to his winning methodology. Through it all, Lou Holtz believes in his team-building system. Lou Holtz believes in being up front with team members and he requires that team members:

1. Do What's Right!
2. Do the Best You Can!
3. Treat Others the Way They Want to Be Treated!

These rules, according to Holtz, answer the three questions asked of every team member:

1. Can I Trust You?
2. Are You Committed?
3. Do You Care About Me?

Holtz brings teams together, as do the leaders of a FlowCycle journey. The team must commit to one vision and support the captain of the ship. In a FlowCycle implementation, coaches facilitate a series of team-building exercises for the Steering Committee at an off-site meeting. Every detail is important. Even rooming together is critical to build trust. As leaders demonstrate their commitment to the journey and their care for other team members, momentum for process improvements will grow into an unstoppable train. The results can be amazing.

At 63, Holtz knows his system works – it just needed some tinkering for South Carolina. As Lou Holtz knows his system works for football, I know the FlowCycle MARI system works for business. It has been working for a long time.

Leaders must learn to facilitate interactive discussions to get people to open up about the hardships they face in life. Leaders motivate people by communicating personal triumphs through adversity that they have personally overcome. It is time to stand up, be passionate, and be accountable to your team.

Cycle Time Reduction must be a journey,
not a program. People are tired of programs and
they can spot one a mile away. Lou Holtz says,
"You try to change their lives and if you change their
value systems, you can change it on the field as well."

When South Carolina won its first two games of the 2000 season against New Mexico State and Georgia, fans stormed the field in celebration. The same level of enthusiasm can be realized in the first days of an improvement journey. Employees are extremely tired of facing barriers to success. Start knocking down those barriers and watch out. A dynamic, energetic mobilization launch following FlowCycle procedures sets a foundation for achieving organizational improvements no one would think possible. All in just a few months.

Management surveys show that leaders tend to take Mobilization lightly. This is a grave error. Leaders must be willing to address all behavioral barriers blocking continuous improvement. Set the standard at World Class, gather the resources needed to balance weaknesses, and commit to the journey. As one Leader said, Effective change management is a journey. Each change presents new circumstances, challenges, and opportunities. Change leaders have the responsibility, opportunity, and privilege to guide others through the change process. Change leaders must develop new attitudes and then demonstrate excitement and enthusiasm for change.

"Sure, it was uncomfortable, at first,
but now, looking back, it was really fun."

At the facility in Green Camp, Ohio, the leaders surprised everyone on the Monday after Thanksgiving. As people arrived from the long holiday weekend, the senior management team welcomed each employee, with music blaring, and signs hung everywhere, generating excitement and demonstrating irrevocable commitment to a new journey. The folks at this facility rarely ever saw senior management and probably thought they were drunk or on drugs. The shock and surprise of these actions broke through barriers and made the employees realize massive change was on the way and this time it was no joke.

Sure, a few employees resisted, but overall these dramatic actions set the stage for what I consider one of the greatest culture changes I have ever witnessed in a company. There was a very tough union environment. The dynamic launch broke the union barrier, and in short order, union leaders became champions of the process. Management leaders followed through on their commitments, and seeing this, union leaders did likewise. This resulted in an implementation of amazing speed and results, and the creation of a great place to work.

The Off-Site Team Building Experience

How is it possible to unite key management around a common vision in a short time? I have found that a three- to five-day off-site training session centered on a series of intense team-building exercises is usually all that is required. These exercises focus management on the importance of the journey and give them the opportunity to begin healing old wounds and open lines of communication. A passive management team will never create enough passion to change an organization. Management must get people's attention and that may involve dynamic actions. The goal is for the Steering Committee to emerge from this off-site experience fired up and committed to the challenges of change.

The team exercises force managers into somewhat uncomfortable situations and require openness and honesty. Why? Because these managers are going to be held accountable for the success of the journey and they must step up to the plate. Leaders must be able to ask their subordinates, "Are you committed to the journey? Can you be trusted to carry out your Mission Task

For rapid change, employees must commit to the journey and trust each other. Team building training teaches needed skills.

Statements? In return for loyalty, I, as your leader, will show love and commitment to you as a person and valued employee. Now let's go out there and show our team we operate under one agenda with everyone supported to succeed and become repeat winners."

As the FlowCycle education process moves forward, all employees at the facility will carry a card of the three FlowCycle principles, four phases, 13 steps, and their leader's vision statement, and a place to sign signaling their commitment.

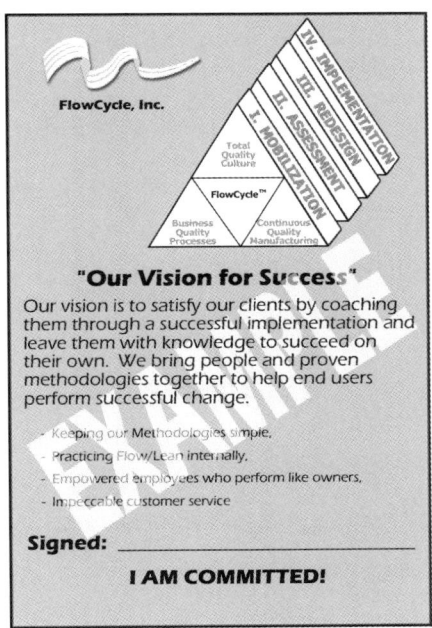

FlowCycle, Inc.

Total Quality Culture

FlowCycle™

Business Quality Processes

Continuous Quality Manufacturing

"Our Vision for Success"

Our vision is to satisfy our clients by coaching them through a successful implementation and leave them with knowledge to succeed on their own. We bring people and proven methodologies together to help end users perform successful change.

- Keeping our Methodologies simple,
- Practicing Flow/Lean internally,
- Empowered employees who perform like owners,
- Impeccable customer service

Signed: _____

I AM COMMITTED!

Phase I. Mobilization
Step 1   Build Foundation for Successful Process Change
Step 2   Select the Processes for Redesign
Step 3   Launch Change Management Efforts

Phase II. Assessment
Step 1   Analyze Customer Requirements
Step 2   Assess Current Performance
Step 3   Conduct Benchmarking
Step 4   Analyze Business for Leverage

Phase III. Redesign
Step 1   Set Design Goals and Priorities
Step 2   Create Further Design
Step 3   Document Pilot for Change

Phase IV. Implementation
Step 1   Develop Detailed Implementation Plans
Step 2   Pilot and Rollout Test
Step 3   Rollout and Continuously Improve

Mobilization Summary

Some leaders may feel that the methods that I have talked about sound soft and fluffy. At my company, I emphasize that these methods are not a social experiment and are in place to produce real change with sustainable results. What I have described may seem like simple concepts, but they are not quick fixes for organizational improvements.

If these concepts were easy to practice, we would see them in more widespread use because of the business and human development success that results from their use. The senior leader and his or her team must drive the MARI system for improvement. It cannot be delegated without changing performance plans. One must have constancy of purpose that comes from the belief that involving everyone in the execution methodology improvement method yields superior bottom-line business results. By permeating this belief throughout the organization, the company will be aligned around one vision to create focus. With this organizational alignment to the vision, the stage will be set to speed the success of Cycle Time Reduction implementation.

- Build a foundation – start at the top by putting management at the bottom of the organization chart.
- Select the critical value streams for improvement based on data.
- Launch change management efforts by taking the risks associated to change and holding people accountable.
- Deliverables:
  - Review and change all performance plans.
  - Show every person where they fit on the team.
  - Build a top-level value stream map of the order fulfillment from customer order through completed payment.
  - Identify new workflows, process families, and potential for elimination of non-profitable product families.
  - Build a Center of Excellence.
  - Build Continuous Quality Manufacturing and Business Quality Process centers and teach quality people how to use them to collect real time data.
  - Take risks and change all performance plans to drive accountability.

CHAPTER FOUR

FlowCycle Phase II
Assessment

| 1. Analyze Customer Requirements | 2. Assess Current Performance | 3. Conduct Benchmarking | 4. Analyze Business for Leverage |

Let's make one thing clear – Assessment of workflow processes involves much more than creating a value stream map on a piece of paper. Getting Lean is not easy. There is a tremendous amount of hard, detailed work and analysis required for the journey to yield positive results. Tracking data, creating visual tools, and analyzing all the information to make Cycle Time Reduction improvement decisions is central to Lean.

Lean is not a cost-reduction edict from top management.
This is why it is so critical to get everyone involved. The job
is bigger than a handful of "experts" can accomplish.

Cost reduction edicts are one of the most destructive, counterproductive mistakes upper management can inflict on a company. These edicts insult the work ethic of every employee and do nothing more than incite retaliation.

Permanent and real cost reduction comes only
through cycle time improvements.

Now that the workforce is mobilized, enthusiastic, and energetic, what are they going to do? They are going to enter Phase II, Assessment to gain a detailed understanding of the current state of their workflows, processes, activities, and tasks.

Assessment has four steps:

1. Analyze Customer Requirements
2. Assess Current Performance
3. Conduct Benchmarking
4. Analyze Business for Leverage

During Assessment, teams will gain a complete understanding of the current-state map of their work area, and they will understand how their process fits into each sector and the overall value stream order fulfillment process.

Often, organizations are broken up into functional silos.

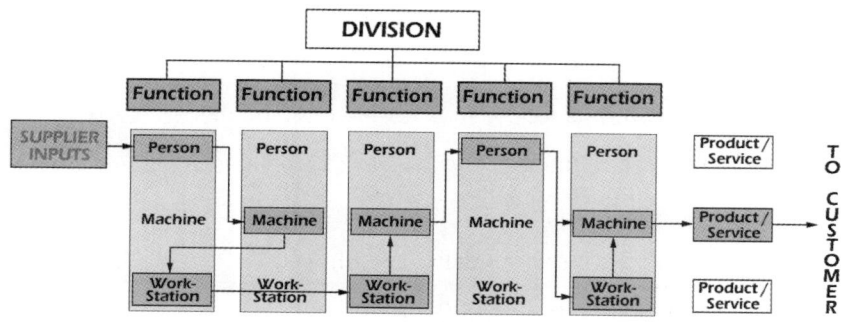

These functional silos feed and/or push information or inventory from one silo to the next. As teams map and assess processes, they are looking for inconsistency or variability. Variability is the consistency of meeting customer requirements.

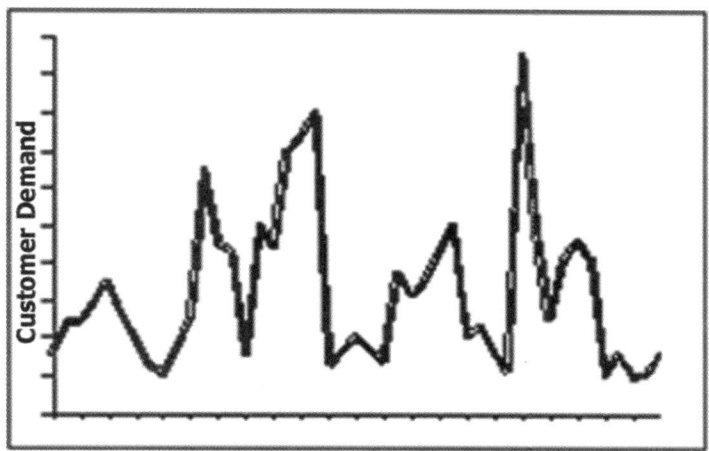

Variability is the spread of measurements which represent process consistency

A good Assessment identifies the strengths and weaknesses of value stream workflow processes. Customer feedback often tells exactly what is wrong. Examples are late deliveries, high pressure sales force, poor product/service quality, and incomplete shipments.

An Assessment should describe the nature and the extent of constraints as well as identify possible solutions. Teams look for patterns to different Assessment questions and then seek to uncover root causes of problems. A pattern of poor on-time delivery performance indicates problems within manufacturing and business processes. Problems with quality and professionalism may indicate a lack of training and too much variability in processes.

> Employees are rarely the cause of process problems.
> Always remember that the process is the problem
> and people naturally want to improve whatever process
> may be problematic because bad processes make
> their job difficult and frustrating.

The Assessment phase uses diagnostic tools to analyze workflow processes. One primary tool is process mapping or value stream mapping. Process maps and the data collected in their preparation provide the detailed information necessary to complete the four steps of Assessment and make effective improvement Redesign decisions for a future state. The maps are integral to adding visibility to workflow processes. They make waste clearly visible.

It is crucial at this point not to allow "analysis paralysis." In a FlowCycle implementation, coaches generally allow two weeks for the Assessment of an individual process. Action is the goal, so building a process map is not the creation of an artistic masterpiece. It is the visualization of the waste in the workflow.

The goal is to remove the non-value-added waste and leave the value-added nuggets of gold. Value stream maps are everchanging and always under construction.

Step 1: Analyze Customer Requirements

To analyze customer requirements, one must have a genuine understanding of the customer. Completion of this simple worksheet helps identify who supplies information or inventory to a process and then who or where the customer is who receives the inventory or information. Use the note section to communicate special circumstances. What is the top-level activity and the three to five top-level tasks of the activity? Who are the task owners?

PROCESS NAME: _____ **OWNER:** _____

SUPPLIER	INPUTS	ACTIVITY	OUTPUTS	CUSTOMER
		Number:		
		Name:		
		Owner:		

NOTES		TASKS	OWNER/OPERATOR

Who Is Your Customer ... Really?

Before teams can genuinely understand customer requirements, they must first be sure of just who the customer is. Most people, no matter where they are in the process, think of the customer as the external customer. This is faulty thinking. Everyone is both a supplier and customer of the previous and next step in the workflow. For example, shipping is the customer of final assembly. An engineering process is the supplier to manufacturing.

The need to understand internal customers and suppliers is just as important as understanding external customers and suppliers. For most organizations, the term "customer" is reserved for external customers–those who buy or consume the organization's products or services. Convincing an organization of the need to do a better job of servicing the external customer is much easier than convincing them of the need to serve the internal customer better. Often it is difficult for people to change their thinking about co-workers as internal customers. Premier customer service means service at every step in the workflow of processing customer requirements.

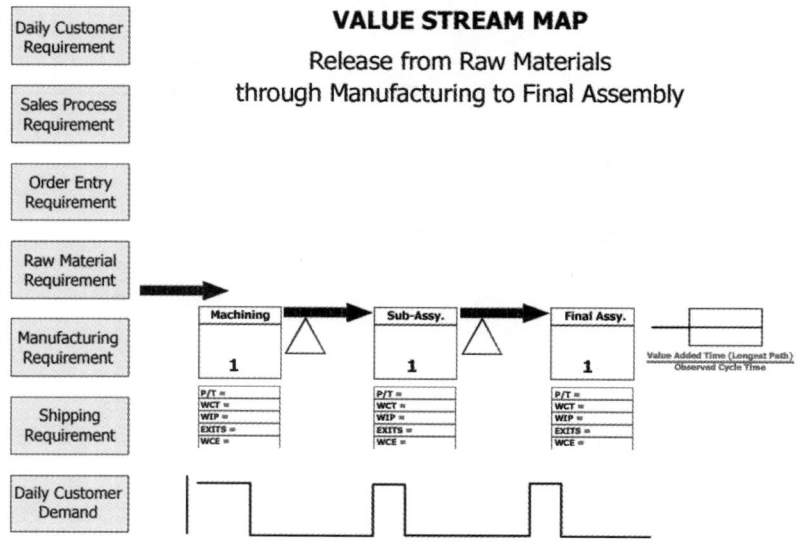

VALUE STREAM MAP

Release from Raw Materials
through Manufacturing to Final Assembly

Surveying Customers

The Assessment phase starts by surveying and analyzing customer data. It is important to remember to survey both external and internal customers.

External Customers

To reduce workflow cycle time, start with the end customer and work back through the process. Sometimes an adequate understanding of the customer already exists as a result of marketing efforts or sales studies. In that case, the Core Team can obtain the necessary information through a team member from marketing or sales. If this is the case, the Core Team should validate that information by confirming it with selected customers. In some cases, the Core Team may decide there is an inadequate understanding of the customers' needs. Tools that the Core Team can use to understand customer needs include surveys, interviews, or invitations to key customers to participate in a roundtable meeting.

Surveys should focus on customer expectations. If the customer reports that delivered service levels do not meet expectations, the survey can discover the customer's expectations. Customers will communicate ideas for improving processes, thereby increasing customer satisfaction.

Measurement data must be collected systematically and quickly to detect changes and trends. It is important to gather sufficient information to make good decisions. The collection of graphically analyzable, multiple series of data is essential.

Measured Customer Satisfaction

Measured customer satisfaction includes development and analysis of measures specific to achieving and maintaining customer satisfaction, customer reject rates, incomplete shipments, incomplete orders, on-time based customer due dates vs. late/promised delivery dates, customer reorder rate, and market share percentage.

Internal Customers

The same survey tools are used for internal customers. Customer satisfaction at every link in the chain is the goal. Surveying internally provides the opportunity to test various survey methods before surveying external customers. Internal surveys aim at discovering bottlenecks, barriers, and dysfunctions, and help teams design smoother workflows. Employees also see these surveys as a trend toward increased professionalism within the organization.

At the factory in the Midwest, many of the hourly and salaried workers did not understand who their customers really were. They continually focused on outside customers. Once they began to focus on both the external and the internal customers, they began to understand what true customer satisfaction means.

At the conclusion of Step 1 of Assessment, teams will be able to define their workflow processes, suppliers, and customers. Teams will understand customer expectations and requirements.

The collected process data includes the number of product variations or part numbers, machines, operations, activities, and interactions. Teams calculate cycle time, changeover or setup times, process times, available machine times, and production times to meet customer demand, production sequence times, the number of operators and/or employees, working times, and number of shifts. Identifying the key output from these activities at each one of the stops along the path, and who the associated customer is at that point, helps identify the needs of input to perform that activity associated from the supplier. Teams then are able to identify the true owner and operator and/or employee for each task.

Step 2: Assess Current Performance

The second step in the Assessment phase is to assess current performance. From Step 1, teams know the requirements of the customer. Now the teams must determine how effective the process is at meeting those requirements. Teams further enhance their process maps in this step by defining tasks as value-added, non-value-added and business value-added. This step assesses the impact of each workflow process activity on the performance measures for that process to identify those activities that add value, those that do not add value, and those that are for internal controls (business value-added).

At the conclusion of this step, teams will have accomplished three things:

1. Constructed detailed value stream process maps that accurately represent the current workflow processes linking each of the sectors together to the high-level map prepared by the Core Team.
2. Collected all the information needed to analyze the process activities and tasks.
3. Formed a value analysis that is visible based on customer value and customer requirements, both internal and external.

Assessing current performance establishes a baseline against which to measure past and future goals determined by current cycle times. The detailed maps break down processes to reveal root causes of constraints and allow for brainstorming for solutions and conceptual design for leverage in future state workflow processes.

Value Stream Map

A good value stream map clearly visualizes the entire workflow and breaks that workflow down to individual tasks and sub-levels. A map must bring visibility to non-value-added activities and allow people to easily understand just what is happening or the map is useless. This is why CQM/BQP centers are one of the keys to sustaining this methodology.

CQM/BQP Centers

Each one of the CQM centers or BQP centers contains information available for review by management or any team member. The following is a list of items that teams present on a center:

CQM or BQP CENTERS

FRONT
- "Can Do's"
- "Need Helps"
- Daily Graphs
- 6 Key Charts

BACK
- Process Issues
- Activity Worksheets
- Process Map
- Transportation Diagram

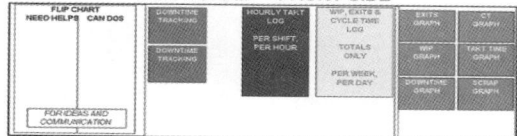

The left side of the front of the center has "Can Do/Need Help" lists of improvement ideas created by the teams.

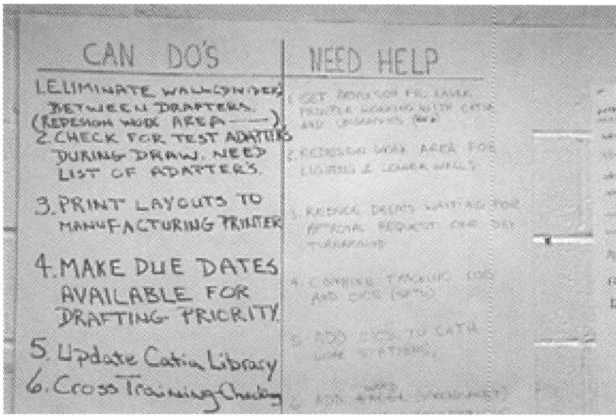

Can Do's are improvements that teams can do, but do not affect the supplier or customer and can be implemented with minimal cost. The Steering Committee sets minimal cost levels, and teams must demonstrate adherence to the MARI methodology. Need Helps are defined as those improvements that are outside the control of a team and the cost parameters and need management sponsorship to accomplish.

The right side of the front center is for visible scheduling.

This is called the Takt board and has the schedule of the weekly demands broken down to daily demand and monitors the actual hourly Takt rate by team.

- The right side arm of the front side of the center has the daily graphs of six key measures: conformance to schedule – on time and complete performance, cycle time, inventory, Takt time – productive time of business or manufacturing process, downtime and scrap.

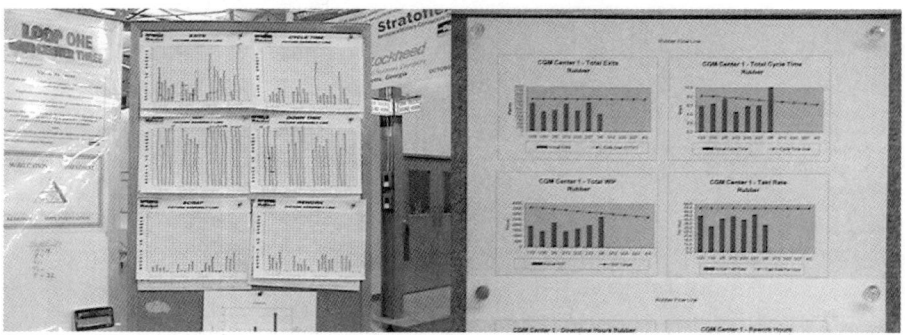

As teams use these centers, they document the process issues, the weekly matrix, activity worksheets and a complete value stream map along with a transportation diagram of their processes. They quickly come to understand how the process flows from supplier to customer.

As teams finish developing their process map, the team calculates the total process timeline based on accumulated cycle times along the value-added path. The current state map shows what is happening right now. The map:

- Identifies wasteful steps in the process and visualizes barriers and delays.
- Visualizes how operations can be combined to a flow-business or repetitive manufacturing production type environment.
- Establishes Kanban/Pull System loops calculated based on cycle times.

- Shows how improvements impact the value stream.
- On the back side of a CQM or BQP process center, the detailed value stream map is tied to the total order fulfillment process map. In the map below, information is coming in from a supplier, order entry and the manufacturing customer is shipping. This simple process map lists the functional areas down the left-hand side, and from left to right is an activity list of major activities occurring in a process.

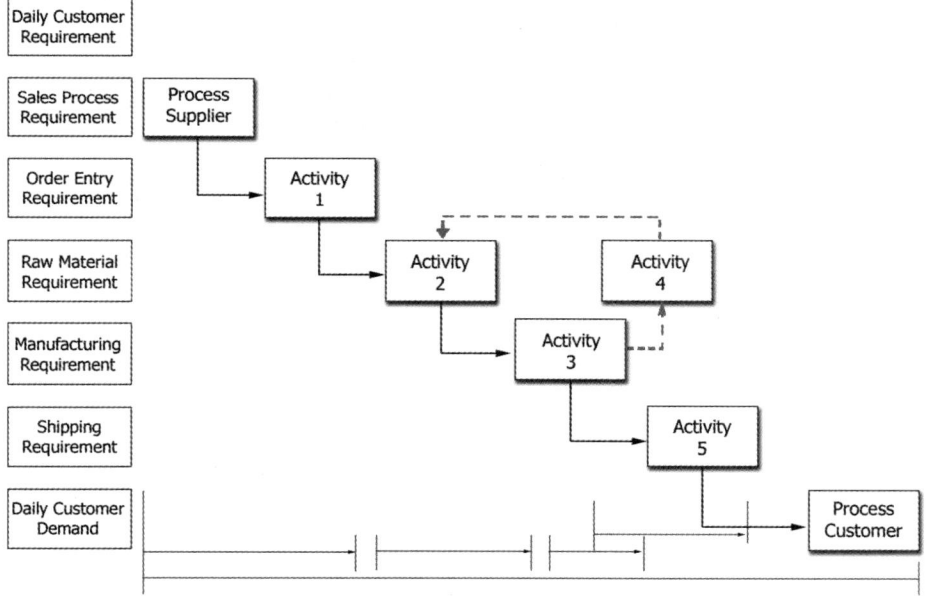

Activities then are further detailed into tasks. The cycle time is the time from the beginning of the first activity of the process until the beginning of the first activity of the next process.

Tasks are sequenced and identified as value-added, non-value-added, or business value-added.

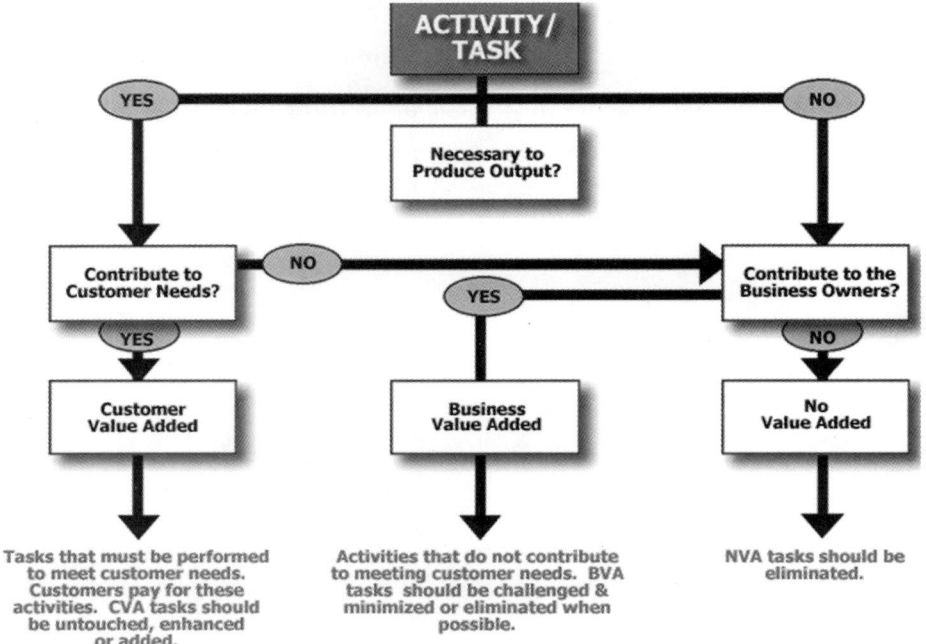

ACTIVITY/TASK		
Tasks that must be performed to meet customer needs. Customers pay for these activities. CVA tasks should be untouched, enhanced or added.	**Activities that do not contribute to meeting customer needs. BVA tasks should be challenged & minimized or eliminated when possible.**	**NVA tasks should be eliminated.**

Developing a process map on the back side of CQM or BQP centers clearly demonstrates the visibility of the process. In a FlowCycle journey, teams use Sticky notes to detail activities and tasks. Blue equals value-added and pink equals both business non-value-added and pure non-value-added. As teams follow this mapping procedure, the charts will have numerous pink sticky notes, and very few blue (value-added). There should be at least 90 percent pink on each one of these flow maps. This clearly illustrates to teams the waste in the process. It becomes fun for the team as they start removing non-value-added tasks.

In the Lean Manufacturing market today, there are many complex mapping processes available that use dozens of symbols to develop comprehensive value stream maps. In a FlowCycle journey, coaches keep it simple by using a mapping technique that captures the multiple paths, decisions, and rework loops in a process without requiring team members to have a Ph.D. in mapping. The following simple symbols denote what each sticky note represents. These five common symbols represent ninety percent of everything that goes on in both a business process and in a manufacturing process workflow.

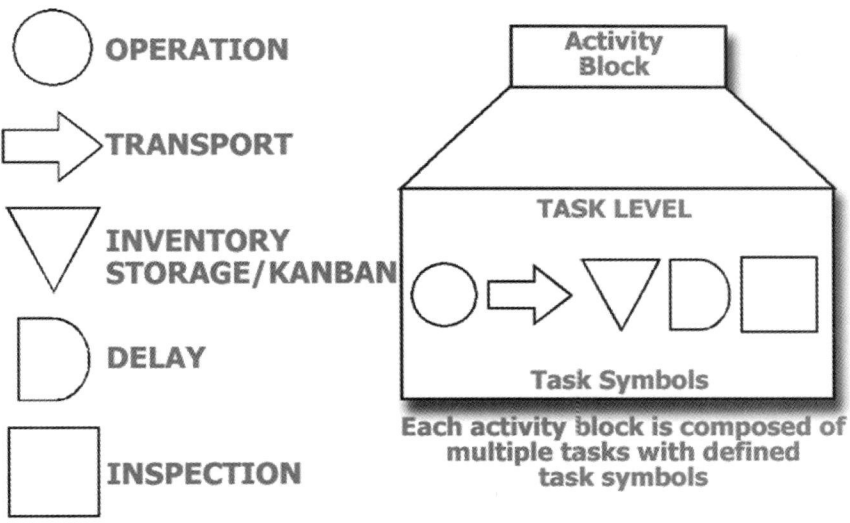

On the sticky note, simply draw the symbol that represents that activity. It could be a circle for an operation and that operation may be value-added or non-value-added. A circle is the only possibility of value-added at any point in time on these visible maps. The remainder of the symbols all represent non-value-added activities.

Activity Worksheet

For quickly developing a value stream map, teams use the following worksheet both in the office and on the factory floor to collect information.

Activity Worksheet

- Types of Activities and Tasks	PROCESS DESCRIPTION			TASK TIMES				

- Value Added
  - Operation
- Non-Value Added
  - Transport
  - Storage/Inventory Level
  - Delays
  - Inspection
  - Business Requirements
	ACTIVITY ORDER	TASK LEVEL	TASK DESCRIPTION	OPERATION	TRANSPORT	STORAGE	DELAY	INSPECTION

			SUB TOTAL	0.0	0.0	0.0	0.0	0.0
			GRAND TOTAL					

**This tool is used in the office or on the factory floor
to collect information on sticky notes.**

The activity worksheet shows each step in the activity with an attached description of the tasks. Then team members check the box that best fits the description of the activity: operation, transportation, storage, delay or inspection.

As the following charts show, each one of the sectors on the process map represents the non-value-added activities. There is only one value-added operation.

Let's back through the process.

- A manufacturing work order activity is released from the order entry department. There is an operation there and also a delay – a non-value-added activity. All activities in this area are non-value-added due to the fact that the product is not altered, therefore the customer would not be willing to pay for those activities.

- Breaking this sector down, teams analyze setups and machine operations. The machine setup has an operation and then a delay, storage and then inspection. The first piece-part inspection occurs before the first true value-added task occurs.

- Once workflows are mapped, the opportunities to reduce cycle time can be pursued. The goal is not to focus on compressing value-added time, but rather to either eliminate the activity or task altogether or perform it in parallel with other tasks so the overall cycle time is reduced. A basic premise of reducing total cycle time is to separate activities between value-added and non-value-added.

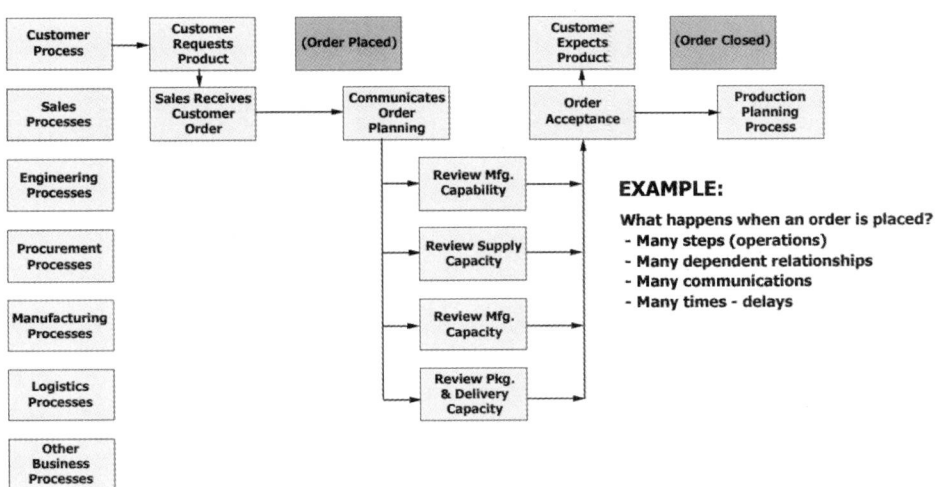

- The process or machine runs once it receives authorization from inspection.

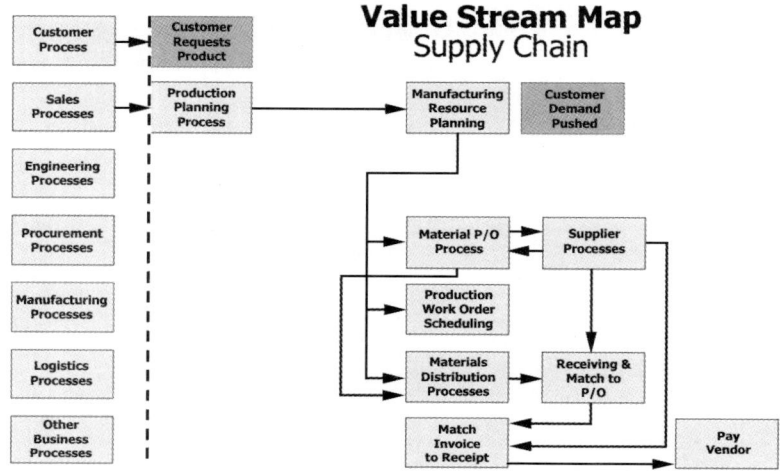

- As business or machine parts are run, they go into a storage location. Extending this approach to the entire supply chain and focusing in on the mainstream activities that add value is key. By providing the output, such as transferring information, from smaller batches much sooner to the subsequent Just-In-Time can compress total cycle time.

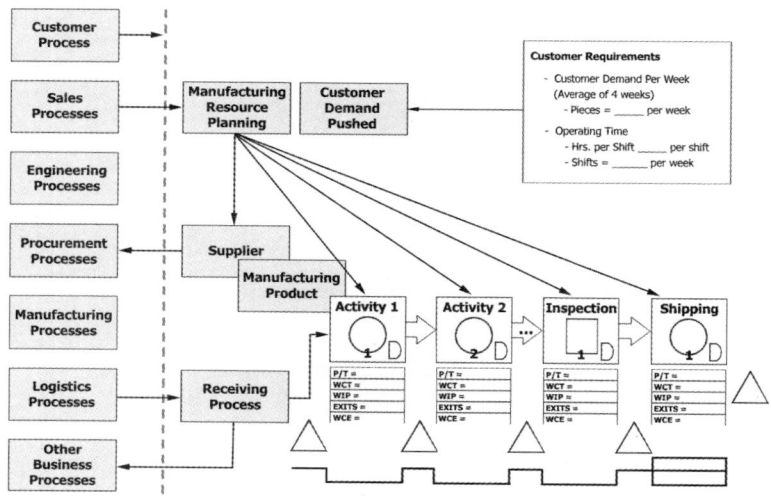

- Then break down processes to the activity levels. Traditionally those activities may be performed in a sequential manner. In this situation each task is completed before the next one begins.

Workstation Process Map

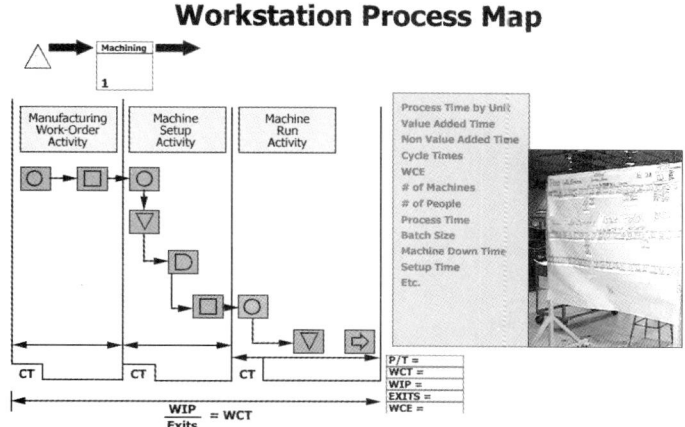

- From storage, the flow moves into another transportation task to feed to the next operation.

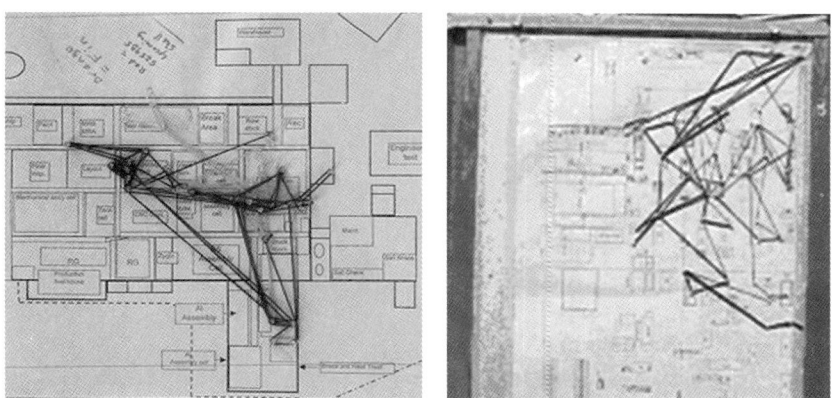

Transportation Diagram - Miles of Waste

Even though this is a very simple process map, there are multiple levels of activities and tasks that happen under an activity block. Often, activity portions of the blocks can be eliminated completely. The map allows teams to see the total cycle time: Inventory (work in process) divided by exits.

Step 3: Conduct Benchmarking

The third step in the assessment phase is benchmarking. Benchmarking is the continuous analysis of the best workflow processes of the industry in order to redesign them. It is a very powerful tool to expand the knowledge of an organization and improve operational efficiencies by measuring the gap between one company's processes and the processes of competitors.

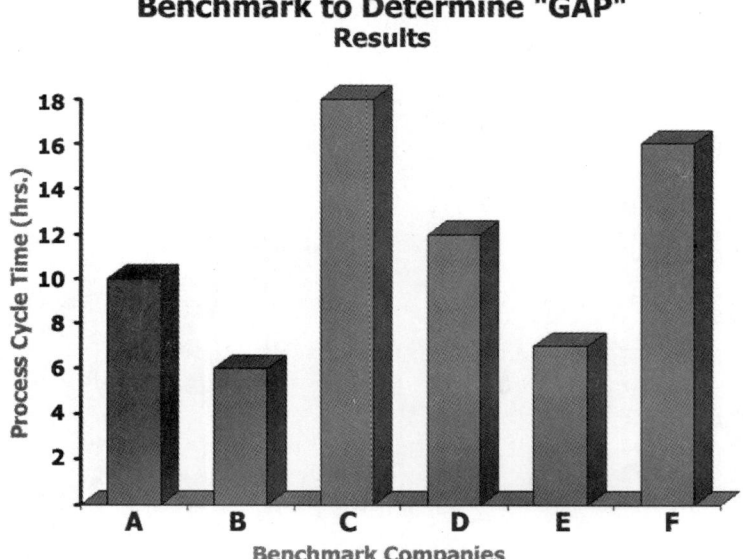

Benchmark to Determine "GAP"
Results

Answer these questions before leaving for on-site:
- What companies perform this business activity better?
- Which company is the best at performing this activity?
- What can we learn from that company?
- Whom should we contact to determine if they are willing to participate in our study?

This analysis is especially helpful in showing ways to improve current workflow efforts as well as revealing important information about competitors. Learning from the outside to validate workflow process goals leads to World Class levels of change. The key is to benchmark organizations or workflow processes that are significantly more advanced. Good benchmarking requires really understanding how all the components of a process fit together. To close the gap with the competition, one must get all the facts and get inside the numbers. Do not just make assumptions, because they waste company time and money.

Look at how many golfers started working out in the gym just because they heard about Tiger Woods' strength. David Duval was the world leader prior to Tiger's rise to the top. Some thought Colin "Monty" Montgomery would be the next Jack Nicklaus. Both went on a weight loss and weight training program in their desire to close the gap between Tiger and themselves. However, body fitness was just one gap between them and Tiger Woods. In the chart below, let's benchmark what Tiger changed. He changed:

- His diet
- His physical workouts before playing every day in competition
- The number of events he entered each year
- His swing and putting
- His mental workouts

Many golfers wrongly assumed physical fitness was the competitive edge, but there were other, less visible but still critical, integrated components.

A workflow value analysis inspects all activities and tasks for their value-added or non-value-added impact. Workflow cycle time analysis is employed for evaluation of positive and negative impacts. Benchmarking is used to quantify existing data and compare them against competitor's practices. If process workflow cycle efficiency is five percent or more, then benchmarking's most important role is to produce new, fresh, and creative ideas for optimizing the next phase, Redesign. This ensures changing those processes that have the greatest impact to customer satisfaction and workflow cycle efficiency.

Benchmarking is a critical step in the Assessment phase because teams now must break paradigms of what they believe is possible to achieve. Often, managers want to purchase a faster or newer piece of equipment to solve a process problem. Sometimes this helps, but most times it does not. Shingo #9 Engine Plant at Toyota runs with some of the oldest engine manufacturing equipment in the world and still outperforms other more modern engine manufacturing plants.

Benchmarking brings attitudes to the surface that can become barriers if not properly addressed. Some barriers that may surface include:

- The fear of loss of competitiveness
- The fear of being viewed as a copycat
- The attitude of operators that they are the best already
- Benchmarking that which is convenient but considered to be unimportant, focusing on core competencies versus the entire value stream to the customer
- People saying, "it will not work," or "ours is different"
- Impatience, which leads to a "do something, do anything" mentality

Benchmarking is not a difficult task to accomplish. First, define the workflows and processes to benchmark. Next, measure your processes in the categories to be compared and select the companies to benchmark. Simply collect data and determine the gap between yours and the best process. Then develop action plans to close that gap.

Benchmarking becomes the means to leveraging change in the workflow process under study. Once an organization has followed the path of Lean for a year or more, benchmarking education becomes a real necessity to enable paradigm shifts for greater improvements.

Good benchmarking practices follow classic FlowCycle methodologies. Benchmarking tends to be best when applied to key workflow processes – those areas that take longer to create change, or those that cause dissatisfaction among customers. Benchmarking is an excellent tool that should be used on a continuous basis.

Benchmarking helps teams understand what is important. It is important to learn from others who have better processes, and then adapt that information to create a learning culture through cycles of learning and cycles of transfer of knowledge.

Once your process becomes the World Class leader, benchmarking becomes a tool to continuously push improvements to widen the lead.

Step 4: Analyze Business for Leverage

The fourth and final stage in the assessment is to analyze business for leverage. Now that teams understand customer requirements, the barriers to maximum performance, and the targets for improvement, they are ready to perform a root cause analysis on those barriers and document a problem statement. Then, Redesign of processes can begin.

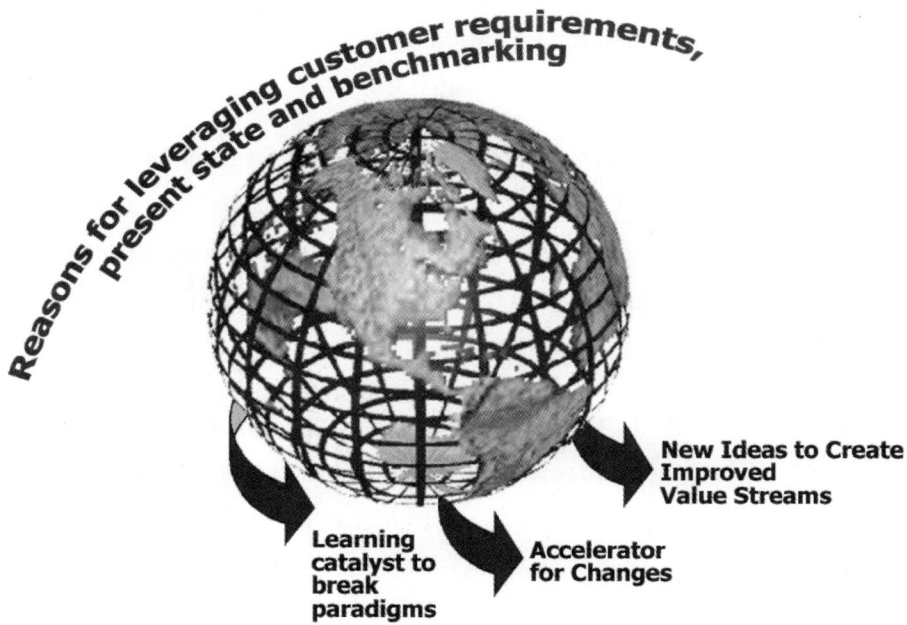

By utilizing the information attained from benchmarking, teams complete a leverage analysis exercise. This pinpoints the greatest gaps between customer perception and performance. This data also identifies ways to reduce those gaps. Many managers simply do not get the performance information needed to make good decisions. This is why it is important that each manager's supervisor lead the effort, identify the data, and then work with this information to improve processes.

Using a Pareto Chart

A Pareto diagram will clearly indicate opportunities for improvement.

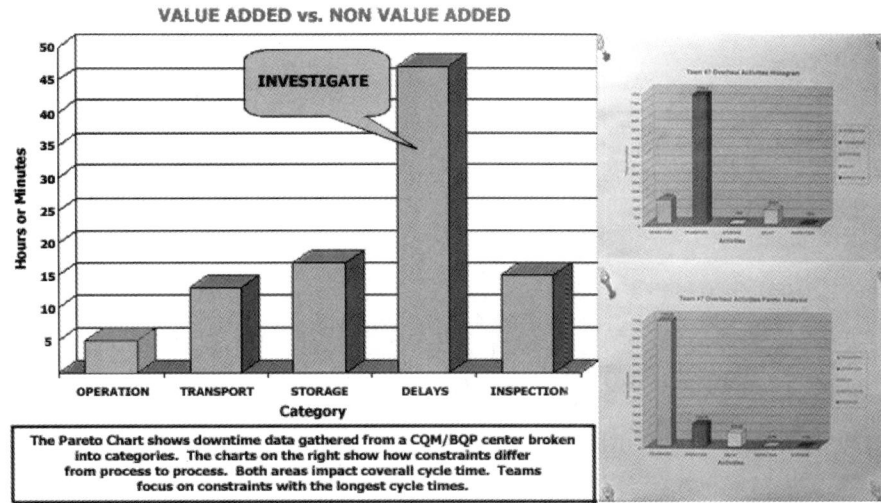

In manufacturing, these diagrams highlight opportunities by analyzing part numbers or product families and signal how to develop product families by pulling pieces of equipment out of the processes. They also reveal low-hanging fruit – easy improvements – through the downtime chart and the transportation diagram.

Managing Resistance During Assessment

During Assessment, managing resistance is critical. Teams must develop a critical path and must understand the deliverables. Often, teams flounder as they begin to try to understand and manage the value stream mapping process. Some team members want to guard information and work on the map until it is perfect. However, the goal of Assessment is visibility, openness, and understanding. Teams must practice the value of trusting and eliminate human anxiety within the culture. As this happens, many people will say, "Why? I told my manager that," or "I told this person that." With a proper Mobilization, ideas previously ignored will surface and improvements in processes and culture will flow easier and more rapidly as barriers fall.

A few years ago, Dover Elevator tried to develop a process in which they could take a very large armature for elevators through the Cycle Time Reduction process. The current process picked up an armature with an overhead crane and set that armature down on a pallet in front of the operation. The armatures built up in inventory before each operation. As the operator began work on the next armature, the first task the operator had to accomplish was to pick little wood pieces from the armature that resulted from setting the armature on the wood pallet. This step had to be completed before starting the wire wrapping operations. This step was completely non-value-added.

As the process maps developed and ideas were discussed, a gentleman came to me and presented a diagram of a mechanical device to move armatures throughout the process.

> The operator had a picture of a device that he had
> designed 27 years before, a cart that would allow stacking
> the armature and rolling it from operation to operation.
> Management's past failure to listen to his good idea had
> cost the company millions of dollars.

Don't make this mistake. The process is the problem and the people are the solution. Comprehensive value stream maps created using simple methods bring the best ideas out of everyone. Many ideas come up through the assessment process. Each idea is valid and needs to be listed as Can Do's or Need Helps on the CQM or BQP boards. What is important is that facilitators and managers listen, hear, and document. People want to know that their ideas matter. Those ideas may never be implemented, but documenting the idea in public means people feel involved. Involvement is the key.

Assessment Summary

- Understand who the customer is – internal and external.
- Understand the customer's real requirement, i.e., cost or on-time deliveries.
- Teach all employees the how and why of process mapping.
- Benchmark to understand what really makes the difference.
- Analyze gaps so teams Redesign the right areas of improvement.

Deliverables:

- Customer surveys
- Value stream mapping
- Analyze leverage by developing Pareto diagram before moving to Redesign

CHAPTER FIVE

FlowCycle Phase III Redesign

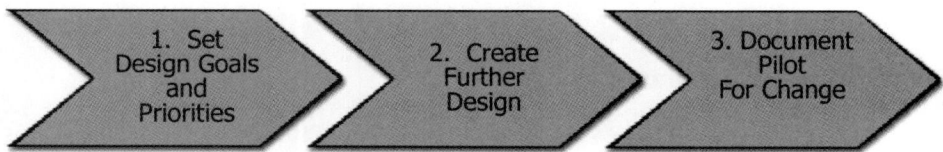

The goal of the Redesign phase is to create and document the future-state designs to accomplish the vision. In Redesign, teams produce workflow designs capable of reducing costs and improving quality for achieving total customer satisfaction. The purpose of this phase is to drive costs down. That's why the triangle is upside down in the introduction of MARI. This phase answers the question "What's new that'll reduce the bottom line?"

The Redesign phase of FlowCycle has three steps:

1. Set design goals and priorities.
2. Create further design.
3. Document and pilot for change.

Based on Pareto diagrams developed in the assessment phase, step 4, Redesign analyzes the process to understand which improvement toolset teams must apply to the process to maximize initial improvements. Teams also analyze data looking to break processes into new, Leaner processes based on product or process families.

Improvement of workflow processes involves eliminating, rearranging, and changing activities and tasks to make processes more effective and efficient. Frequently, process change depends on the team seeing the pink, the non-value-added-tasks, to prevent quality defects at the source, banish culture barriers, correct workflow processes, and create new designs. As teams gain experience, they begin to see more and more waste, and they also begin understanding how improvement tools apply in specific situations. They must understand the barriers in both organization and operations that prohibit a company from meeting the cost benefits and exceeding customer requirements.

Over the last millennium, process engineers have designed an entire repertoire of improvement tools to facilitate process improvements. These toolsets should not be viewed as a separate discipline, but simply as enablers to achieve reductions of workflow cycle time and new process workflow designs.

FlowCycle's MARI process places these toolsets in the proper context for application. The list of improvement toolsets designed to help achieve cycle time improvements include the following:

- Team buildings – cross training
- Looking at best practice standardizations
- Process simplification and compression
- Defect prevention – each operator self-checking and self-stopping
- Six sigma
- Set up reduction change over improvement
- Process flow improvements-cells-flow lines-one piece flow
- Bureaucracy busting – eliminating task interference
- Visual management systems
- Group technology in the business processes
- Kanban/Pull Systems in the manufacturing and business processes
- Customer-focused response policies in the process

Step 1: Set Design Goals and Priorities

The first step of Redesign is to sequentially prioritize workflow processes by analyzing the value stream map for product families or processes and services that can be leveraged for change to improve performance. Prioritizing workflow processes intended for improvement has been an integral part of the MARI phases since early in Mobilization. The final step of Assessment points teams to an understanding of which process improvement tool is required based on the gaps identified in the overall value stream and in the individual sectors. The Center of Excellence has been tracking the metrics at the company, division, plant, and sector level, determining the baseline for improvement. The first step in setting design goals is to establish and understand how each is contributing to the overall performance of on-time redelivers and cycle time of company. They may not be the bottleneck or constraint in the company, but remember, the goal is to drive the cost of quality down in all areas of the company.

Setting Design Goals

Premier customer service is meeting requirements as defined by the customer. Creating new goals for the six was agreed on at the needs assessment. The benefit of process changes, tools, organizational changes, and so on cannot be accurately assessed without metrics. Obviously, the goal is to tie premier customer service in the five supporting metrics, and many teams start improvement efforts at that point with the goal of reducing cycle time by fifty percent. We must not stop with this goal – cycle time can only be reduced two ways. Doubling output or reducing WIP or inventory by half will not create improvements. Why? Because the Pareto charts were describing the drivers of the current cycle time. Make the team understand relevant metrics that help them actually improve in ways that help workflows. Team members may not agree on things as simple as a new layout when the project starts, but they will agree on the need for change when well-defined metrics they collected bring them to that point. Measures must drive the following:

- Eliminate the constraint.
- Reduce costs.
- Sequentially institute ownership accountability and a standardization of the process functions and tasks.
- Measure the process performance.
- Tie performance to the appraisals and compensate people for the proper behavior.

Use quantified data, not seat-of-the-pants guessing, as a basis for redesign improvements. Avoid "but not invented here" attitudes by including everyone on a Redesign team in identifying, communicating, and utilizing reliable methods that strive for continuous improvement. Teams examine performance criteria, financial criteria, information technology criteria, implementation criteria, and most important, human resource criteria for success.

Setting Priorities

Now, the Core Team and Steering Committee use all the data compiled in Mobilization and Assessment in identifying, mapping, and analyzing processes. These processes establish the order fulfillment processes and will be improved by designing a Kanban/Pull System. The second priority is determined if there is enough resources of people, machines, and floor space to design product or process family flow cells.

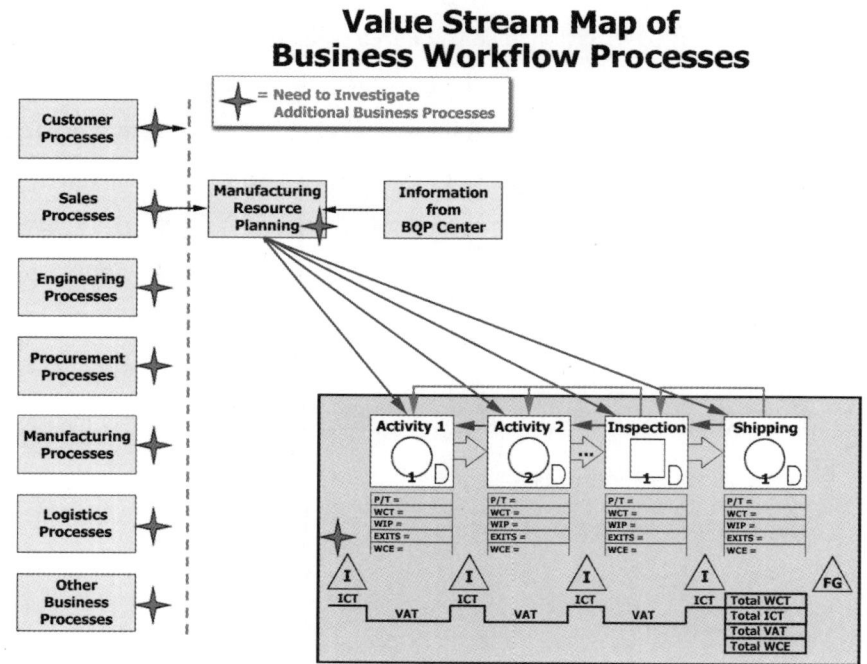

Value Stream Map of Business Workflow Processes

Implementation teams are focused on removing the low-hanging fruit that was identified in their centers as "Can-Do's" from value stream maps. Implementation teams should follow the same Redesign and implementation step to capture the improvement implemented and be given credit for their results. Implementation teams should prioritize based on the simple changes in profiling each Can-Do and documenting the savings. All teams should follow the three key areas of setting design.

Redesign Check Off

		PROCESS			ORGANIZATION					TECHNOLOGY		
Root Causes	Alternative Solutions	Policy	Practices & Methods	Process Flows	Organ. Structure	Job Redesign	Physical Layout	Training	Mgmt. System	New Technology	Appl. Systems	Relations w/Vendors

Determine which process improvement workshop must be implemented to eliminate the constraint. (The chart below indicates a workshop in the area of setup reduction is needed to reduce the batch size, which will ultimately reduce cycle time.)

Building a House That is Supported for Long Term Growth

WORLD CLASS PERFORMANCE

COACH DEVELOPMENT WORKSHOP

SETUP REDUCTION WORKSHOP

PROCESS FLOW IMPROVEMENT WORKSHOP

OPERATION TIME IMPROVEMENT WORKSHOP

TOTAL PRODUCTIVE MAINTENANCE

DEFECT PREVENTION WORKSHOP

WORKFLOW IMPROVEMENT SKILLS MODULES

KANBAN/PULL SYSTEM DISCIPLINE

ORGANIZATION/LEADERSHIP DEVELOPMENT

At the bottom of the pillars, organizational skill and leadership development improvements began on the first day of Mobilization.

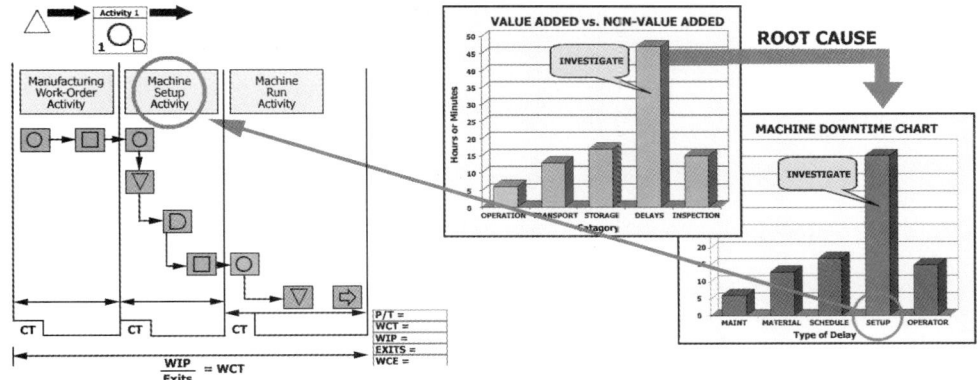

A Kanban/Pull System, (see step 2) brings stability and discipline to the process. At the bottom of the pillars, the organization and leadership skill development is critical to the sustainability of any improvement process. This has been identified from the first day of need assessment into Mobilization. A Kanban/Pull System discipline brings stability to the process. Finally, teams apply process improvement toolsets like setup reduction, defect prevention, operational time improvement, total productive maintenance, etc., to improve the value stream.

Step 2: Create Further Redesign

In Redesign, step 2 – Create Further Redesign, teams brainstorm for creative ideas based on the improvement workshop tool selected by the team in step 1. If the number-one constraint was setup reduction, the team would create further Redesign based on training in that module. There is never only one solution with any improvement workshop, so everyone's opinions and ideas are valid for consideration and analysis. Team leaders must manage all these ideas. Team members will have their own opinions, and those opinions need to be managed.

Brainstorming sessions require the following ground rules and guidelines:

- Who will participate and why
- Outside-the-box thinking not simply what is "good for me"
- Short-term and long-term ideas
- Challenging conventional wisdom
- No idea is a bad idea at this point

I had the opportunity twice in my life to meet Dr. Deming. The first time was in an open forum in Orlando, Florida. During a speech to IBM employees, he asked for questions from the audience. I will never forget what happened. A young lady walked over to the edge of the stage to a microphone and asked Dr. Deming a question. Dr. Deming replied that her question was the most stupid question that had ever been asked. I was astounded. This was supposed to be a man of quality. His response taught me a lesson.

Ideas should be challenged, not people.

Teams must become engrossed in the improvement process. Leaders must encourage fun, openess, and trust and be willing to chart new waters. Bad ideas and "stupid" questions are integral to the learning process.

A basic assumption of dramatic workflow process improvements is that the activities and their supporting technologies can be supported and designed in parallel. The result of the workflow process design effort includes a complete definition of technological support requirements for the organization.

These requirements include the technology required to assure the success of workflow process change. Teams should include Information Systems personnel on the team, or have clear channels of communication with IS to communicate design requirements. Teams will send designs over to IS with much, if not all, of the system analysis and design completed. The redesign may require an investment in software or the creation of some kind of new user-friendly platform.

Net Requirements

The goal for overall Cycle Time Reduction starts at fifty percent, minimum. This is managed through the first phase of a Kanban/Pull System called net requirements. Net requirements is a methodology that allows the Core Team to limit the amount of inventory carried in Kanban/Pull point-of-use locations, supermarkets, buffers, and safety stock. Net requirements are the calculation of the amount of inventory required to service real customer demand.

In fact, the inventory reductions usually pay for the project within the first three to six months of implementation.

Teams calculate net requirements by analyzing the following:

- A key element of the future-state map is the definition of a plan to reduce inventory in raw materials, WIP, finished goods, and service parts. This plan should address the tasks required to move from the current situation to an eventual system that reduces costs and delivers product to the customer when required. Remember, product can be an engineering drawing, work order, or release order from quality.
- Process flows from supplier to customer internally on the transportation diagram indicate where net requirement levels should be set in the Kanban/Pull locations.

Developing Future State Value Stream Map
Kanban/Pull System

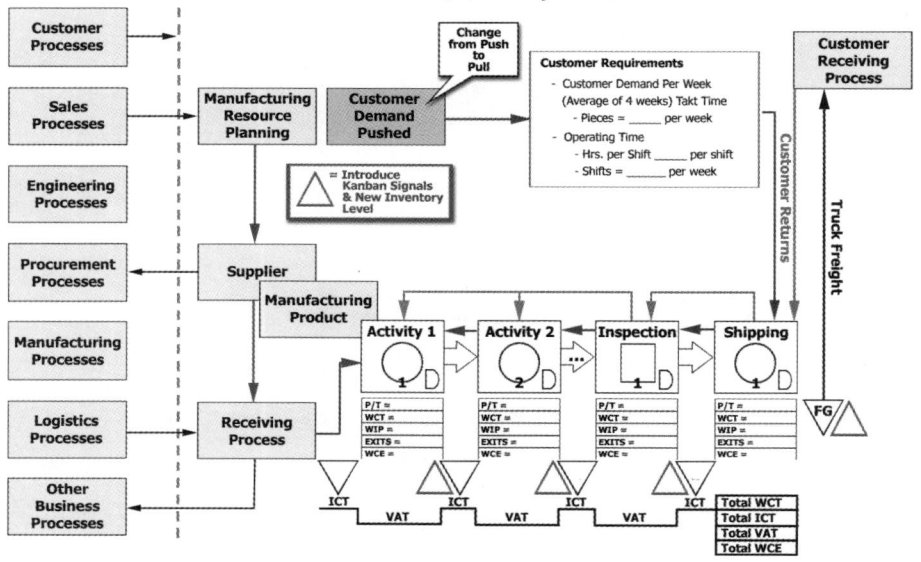

The Core Team will create a future-state map that shows the relationship between achieving the new metric of 50% Cycle Time reduction and increased throughput. The Steering Committee and Core Teams must compare the current-state maps to future-state maps to ensure during the implementation of the new workflow that there is enough inventory to meet customer requirements. The future-state map should understand the relationship between throughput requirements for proper design of Kanban/Pull levels to meet the Takt rates of the customer. Kanban/Pull calculations into changing environment are dynamic and must be reviewed at least monthly if not weekly, because customer demand is changing much more frequently. I look at Kanbans as the trigger to create a Pull System. They are simply the triggers that are based on the customer demand, cycle time, cycle time intervals, and safety stock – (CD x CT x CTI + SS). When designed correctly, there are enough inventories or works in process to draw from to meet the Takt rates, which at this point have no conflict with absorption.

The Takt rate will set the flow rate being pulled from each operation to meet customer schedules internally and externally. To be able to determine the proper cycle time, the processes must be managed to a Takt rate – the heartbeat or rhythm of the process.

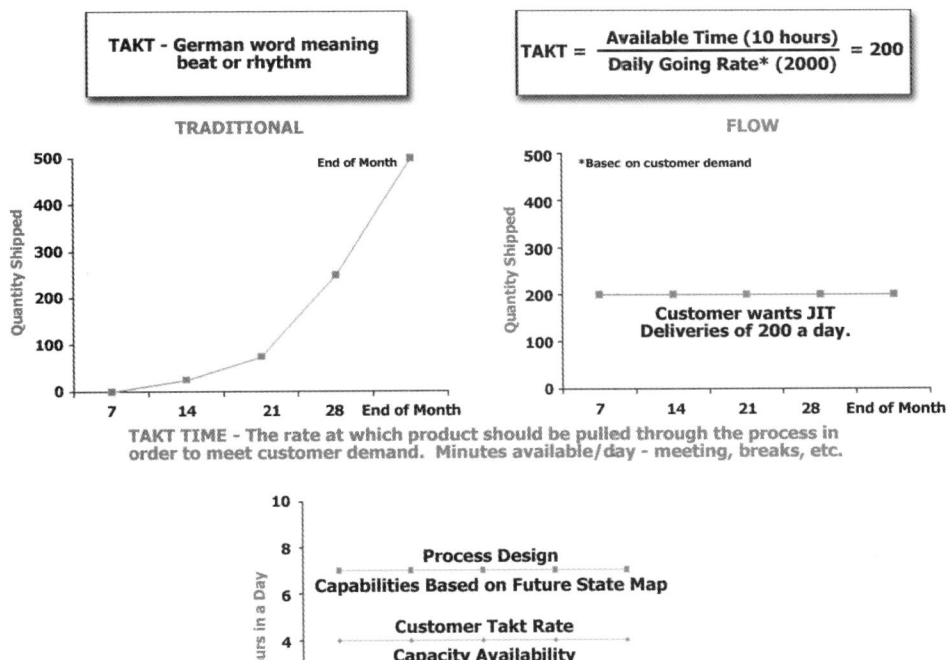

The goal is inventory built on a linear basis and shipped to customer demands on an hourly/daily/weekly basis. Takt rate is equal to the available time in a day minus meetings and breaks divided by the daily shipping rate.

Net requirements also set the priority for work to be completed to meet customer due dates. This concept prevents cherry-picking of jobs by operators and can be applied in the business processes as well as manufacturing. People in all areas of a company tend to cherry-pick tasks, which ultimately creates the need for heroes and expediters to overcome the problem of tasks done out of sequence.

PRIORITY OF WORK SCHEDULING

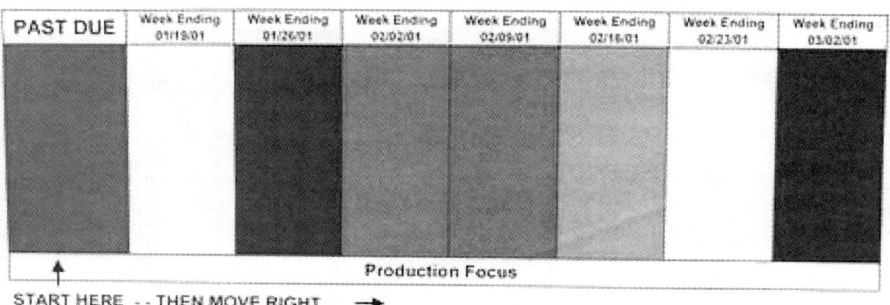

PAST DUE	Week Ending 01/19/01	Week Ending 01/26/01	Week Ending 02/02/01	Week Ending 02/09/01	Week Ending 02/16/01	Week Ending 02/23/01	Week Ending 03/02/01

Production Focus

START HERE -- THEN MOVE RIGHT →

Shop Orders are color- coded per the above schedule using the Master
Schedule to determine correct color.

Often, more than 95% of the order-to-delivery cycle time consists of waiting due to cherry-picking. This not only happens in manufacturing from sales orders, from engineering documentation, from corrective actions, from IS programming, but it just happens to be human nature. All of this waiting keeps stretching the order-to-delivery cycle time while adding unnecessary operating expense in the firefighting process to correct the problems that should never have been created in the first place. It's a vicious circle that feeds on itself, and one that should be stopped. Fortunately, net requirement has a simple visual tool that is comparatively easy to accomplish and offers numerous benefits. The number-one benefit is it visually shows workflow on right or wrong customer orders in all areas of the company.

Step 3: Document Pilot for Change

In Redesign step 3 – Document Pilot for Change, teams create their future-state process map. Starting with the current-state maps created in Mobilization and Assessment, teams summarize strategies for new technology, summarize the Redesign cost and benefits, package and communicate the new solutions, and develop plans to remove barriers based on communication. This data develops a business case for change and a future-state map.

The new future-state map may drive organization changes, and teams must make these changes visible on a new organizational chart. The following current- and future-state maps illustrate the application of a Kanban/Pull System and the moving of inspection in line.

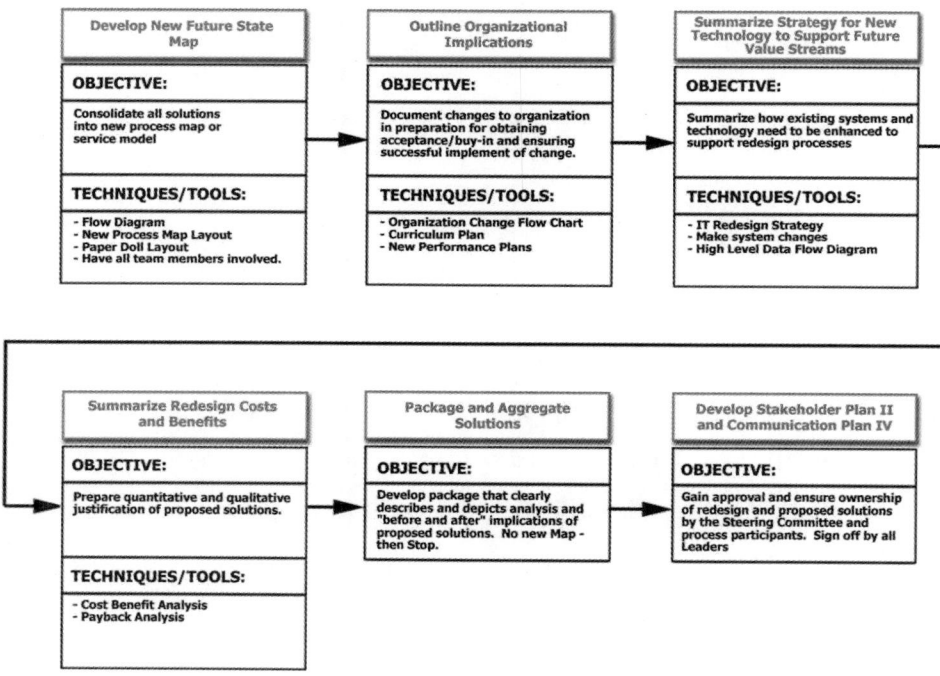

Teams examine organizational changes and analyze the relationships and responsibilities of who is reporting to whom. Teams apply the future-state mapping process to any functional flow for Cycle Time Reduction. For example, the maps below illustrate current and future human resources workflows. What are the staffing level requirements? What are the field and training requirements needed? What does the measurement and reward system look like?

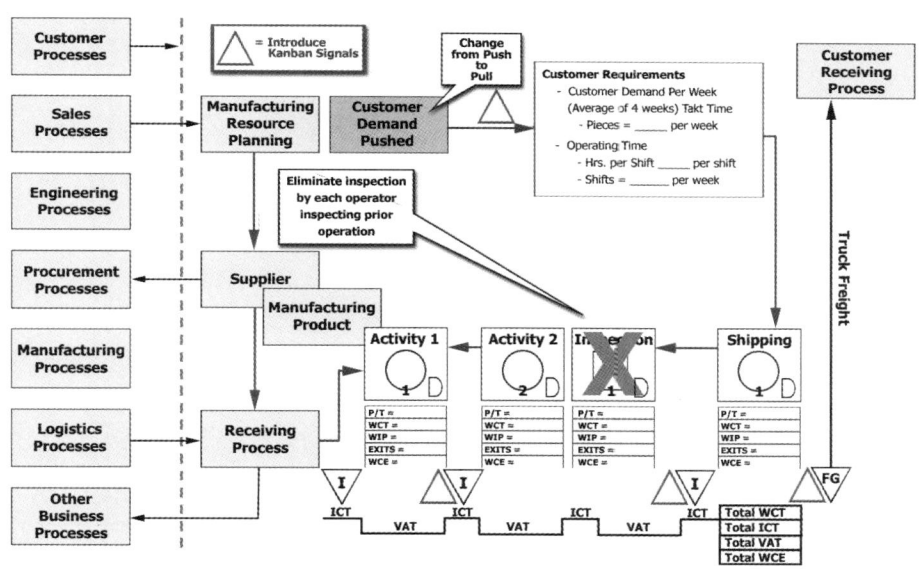

The most important outcome of net requirements and the first level of implementation of Kanban triggers is it will free up space that can be used for future-state designs of new product family layout and/or moving product families into an exciting facility with the cost of land, brick and mortar cost. The factory in Brazil had to move due to government requirements. When I started the project with them, they wanted to design a building approximately the same size as the old. The future-state map of all the product families convinced them to build a new building 1/3 the size of the original.

They followed through and reduced the factory from 800,000 sq. ft. to 270,000 sq. ft. on a greenfield site. The facility was built and all value stream product families were in place in nine months. This is why the last step of Redesign is so important. Documentation is the key to success.

The following is an example of a process flow layout.

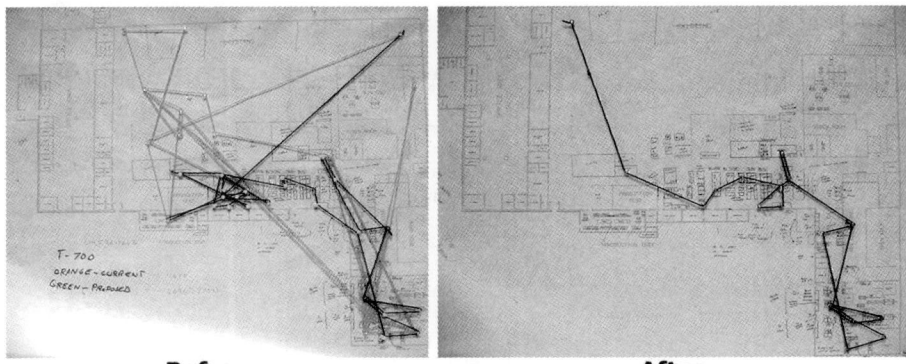

Before   **After**

Performance Plans

The final element of Redesign is to redesign performance plans so that employees' behaviors change along with the new system. There seems to be reluctance in many organizations to change performance plans to center around Cycle Time Reduction. Companies too often change the workflow processes without changing the assumptions about personnel management issues. People work on those tasks against which they are measured. Managers assume that it is impossible to get 100 percent improvement on performance from either the workforce or an individual on a consistent basis. These assumptions are not correct and must be challenged. In order to improve continuously, individuals must have incentives tied to improvement.

There are dangers lurking in this step of the process, so managers must carefully consider their actions and always include human resources personnel. For example, if managers display "get tough" attitudes and demonstrate too much enthusiasm, especially in their willingness to fire employees, they can quickly lose credibility and trust and the entire journey becomes imperiled as some team members may attempt to sabotage and thwart process improvements. This is why human resources must be involved on the Core Team to ensure performance plans are fair and accurate in defining behaviors.

HR's involvement is of great importance to ensure that change is institutionalized; that is, a continuous improvement culture becomes the norm. In recent years, many continuous improvement initiatives have stalled because of a failure to align performance plans with the journey. The overall recommendation for developing performance plans is to change them from the top down, and implement them from the bottom up.

Frederick Taylor, the father of the industrial engineering effort that created the word "Reengineering," reinforced the view of the workforce as part of the work-flow processes. He taught that the redesigning of processes is the responsibility of industrial engineers. Whatever the mechanical or process engineer created, the industrial engineer had the responsibility to improve. Leaders must apply this same principle to all employees. There should be some-thing within every employee's performance plan that guides him or her to be a "mini industrial engineer," with the responsibility to continually improve his or her processes. The ideal of workflow process improvement and continuous improvement, with Cycle Time Reduction as the key driver, is that every manager and supervisor buys into the process with complete trust and cooperation.

> The only way to guarantee buy-in is to include Cycle Time
> Reduction as a key component of performance plans.

Organization Structure and Responsibilities

1. Structure Supports Strategy

Work and resources are organized so that:

- They focus on customer deliverables
- Roles and responsibilities in the provider/ customer continuum are clearly defined
- Each work unit has the kind and the amount of resources necessary to serve its customers (both internal and external)

By facilitating decision making and responsiveness:

- Structure minimizes the need for inter-segment coordination and decision making
- Management levels are minimized, shortening communication channels
- Decision making is pushed closer to the point of customer contact
- Structure is readily available to changes in:
    - Market situation
    - Business conditions
    - Technology
    - Competition

Staffing Levels

2. Management positions exist to facilitate the work of those who serve the customer

Managers assume responsibility for:
- Vision
- Mission Task Statements
- Strategic direction
- Coordination of Improvement Efforts
- Develop Resources and Coaches
- Skills Development Requirements
- Empowerment and Accountability

Position and layers are kept to a minimum, so as not to:
- Block communication
- Confuse decision making
- Add unnecessary work approval layers

3. Jobs/ processes are designed to empower the employees to serve customers

- Activities are naturally related, so that people can see the beginning and end of their work and its relationship to the company vision
- People are able to assess, manage, and improve their own performance
- Each job is a complete element in the continuous cycle of creating and serving customers; dependent activities are linked to value streams
- Each workflow affords opportunities for skill and career path development
- Value Streams are clearly linked into workflow process that have clear outputs valued by your customers

Skills and Training

4. Key Elements of a Curriculum Plan

- Skill Curriculum
- Workshop Contents
- Course development resources
- Prerequisites
- Instructional strategy
- Delivery strategy/what/when
- Course Outlines
  - Executive/Management 2-Day Course
  - 5-Day FlowCycle Fundamentals
  - M.A.R.I.

- Workshop - course time requirements
- Training facilities and location
- Evaluation strategy
- Workshop curriculum map
- Workshop descriptions
- Implementation classroom schedules
- Understanding of work assignments
  away from classroom

Measurement and Reward System

5. Total Rewards Strategy	6. Group/Team Incentives	7. Individual Rewards Strategy

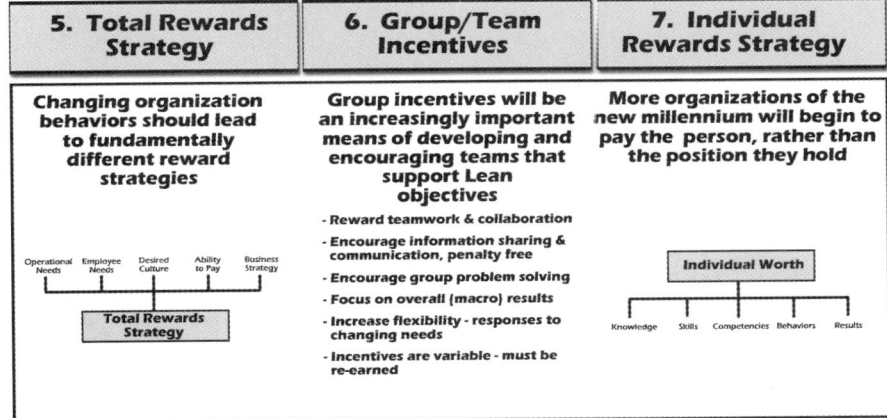

5. Total Rewards Strategy

Changing organization behaviors should lead to fundamentally different reward strategies

Operational Needs · Employee Needs · Desired Culture · Ability to Pay · Business Strategy → **Total Rewards Strategy**

6. Group/Team Incentives

Group incentives will be an increasingly important means of developing and encouraging teams that support Lean objectives

- Reward teamwork & collaboration
- Encourage information sharing & communication, penalty free
- Encourage group problem solving
- Focus on overall (macro) results
- Increase flexibility - responses to changing needs
- Incentives are variable - must be re-earned

7. Individual Rewards Strategy

More organizations of the new millennium will begin to pay the person, rather than the position they hold

Individual Worth

Knowledge · Skills · Competencies · Behaviors · Results

With new performance plans in place, organizational development is complete. Personnel issues or conflicts in the organization have been removed, and everyone is aligned for rapid progress on the journey. Implementing process improvements can proceed uninhibited.

Redesign Summary

- Designing new goals of at least 50 percent reduction of cycle time in all areas.
- Understand customer priorities.
- Document and pilot.

Deliverables:

- Posting of new measures throughout company.
- New Performance Plans based on Cycle Time Reduction.
- Roadmap with conceptual designs, analyses, simulations, and economic benefits that communicate the expected results before implementation begins.
- Preliminary design to show how the future-state map will meet the strategic objectives.
- Simulation that demonstrates future-state performance. This simulation also serves as a very effective tool for communicating plans and training employees prior to implementation.

CHAPTER SIX

FlowCycle Phase IV Implementation

| 1. Develop Detailed Implementation Plans | 2. Pilot and Rollout Test | 3. Rollout and Continuously Improve |

The Mobilization, Assessment, and Redesign phases of FlowCycle are all preparation for actually implementing workflow process improvements. Teams have done their homework and redesigned their processes. Now it is time to plan, pilot, and implement improvements and continuously cycle through the four phases and thirteen steps of MARI. Implementation should be seen as only the beginning even though it is the nail in the coffin of old work habits and old workflow processes. The Implementation phase is the start of continuous improvement. Implementation is change made real.

Company improvement teams are often excellent at Assessment and Redesign, but fall flat at Implementation. The reason? They forget Mobilization and they try to "shoot from the hip" in Implementation instead of meticulously and systematically continuing to apply the journey plan.

Buy-in is essential during Implementation. The process of getting buy-in began on the first day of Mobilization, so actually achieving buy-in for an improvement should be easy. Nevertheless, buy-in cannot be assumed.

There are three key steps to implementing workflow process improvements with speed:
1. Develop detailed implementation plans
2. Pilot and rollout test
3. Rollout and continuously improve

Training during Implementation is important. Team members want to know how the new process operates, their particular roles and responsibilities, and how new measures affect them. Piloting the Implementation on a small scale helps eliminate uncertainty and identify unforeseen barriers.

Step 1: Develop Detailed Implementation Plans

WHY
- To determine what to change, what to change to and how
  Include all activities and steps of value streams

HOW
- Identify value streams to implement change
- Estimate duration and completion dates for each action step to develop feasible timeline and estimate critical path to achieving new value streams
- Assign responsibilities to process owner, cross-functionally for measurable deliverables and hold people accountable
- Develop milestones to assure timeline synchronization with the critical path with daily action items
- Develop risk-management strategy to optimize schedule attainment

KEY TOOLS (Visible in Center of Excellence room)
- **Planning Worksheet** - actions, owners, due dates, costs
- **Network Diagram** - accountable to completing actions
- **Gantt Chart** - a detailed project plan of daily tasks
- **Milestone Summary** - daily what got done, what was missed, and what recover schedule
- **Risk-Management Planning Worksheet** - customer schedule risks, cost, overruns, risk of key people leaving

Developing and using the right tools to ensure a successful Implementation is the key to gaining momentum in the process. Teams will review workable alternatives, develop plans to gain support of key players, and develop an Implementation plan, all while utilizing the tools in this section for fail-safe and successful implementation of solutions.

A good plan is the first step to successful Implementation. Leaders must:

- Involve the entire team in planning by updating and informing key players of the plan development.
- Keep the plan simple for achieving the desired result. Complexity breeds confusion. The implementation plan must remain simple and be based on common sense.
- Communicate the plan to key players using visual methods, such as a network diagram that clearly shows team members where they fit in the new process. It should also outline each task and its parallel or sequential task, the duration of time, and the person or resource responsible for that task.

- Provide an overview of the developing Implementation plans in the Center of Excellence for management to review.
- Find the critical path for Implementation. Activities and delays at any level need to be brought to the Steering Committee for immediate resolution.
- Solidify senior management support and sponsorship through commitment.
- Review the highest priorities of the solutions and develop a diagram of required actions that shows a responsibility matrix.
- Develop a Gantt chart so the team can visualize the timeline for Implementation and determine the actions to accomplish. MS Project™ is a very good tool for displaying Gantt charts. The template below is recommended for daily use to ensure Implementation action. Tasks are focused through visible accountability.

Task Number	Improvement Action & Plan	Action Steps	Start Date	Completion Planned	Actual	Function Owner	Implement. Cost Planned	Actual	Savings Planned	Actual	Status Open/Close
BENEFITS & DELIVERABLES:					TOTALS						
					NET SAVINGS						

Planning is the stage of the methodology that is, in many ways, the most critical portion of any Implementation. It represents the point of buy-in and is the most involved stage of the process. Planning lays the foundation for all that is to come if done correctly. Regardless of how straightforward or complex the requirements of the Cycle Time Reduction project, the quality of the preparatory work and start-up has a direct bearing on the effectiveness and ease with which implementation will be managed. It is at this point that understanding and use of daily project management and audit tools first come into play.

Step 2: Pilot and Rollout Test

The second step of Implementation is to pilot and test the improvement by defining new performance measurements. Teams select a small area where a pilot can be developed and implemented. The new process is measured and compared to the current workflow processes. Pilots demonstrate success, confirm designs, and guard against wholesale failures and are essential for generating enthusiastic buy-in to the new workflow process improvements.

After running a pilot of the new design, the team makes any adjustments needed and gets buy-in from all the key players based on success. Pilot results are used for presentation development and in educating the team on the new process. For example, teams may use cardboard models of equipment or office furniture to walk through the new process and visualize it.

This is the point when the Implementation team creates daily lines of accountability and implements their solutions. By the time this phase is completed, a blueprint exists for FlowCycle Implementation and all future Implementation activities. Thorough follow-through is the most important characteristic of successful Cycle Time Reduction Implementation. Implementation leaders demonstrate persistence and commitment during follow-up by encouraging team action and removing obstacles.

Answer each question below with "Yes" (Y), "No" (N), or "Unknown (?). When a (N) or (?) answer is given, the team should agree on the action to be taken.

IMPLEMENTATION OBJECTIVE	ACTION REQUIRED:
1. Set Improvement Goal	
___ Is it clearly stated?	_____
___ Is it measurable?	_____
___ Clearly tied to customer satisfaction issue?	_____
2. Identify Problems	
___ Process clearly documented?	_____
___ Is this the most important problem?	_____
3. Analyze and Identify Cause	
___ Cause clearly identified?	_____
___ Cause factually verified?	_____
4. Generate Alternatives/Select Best Option	
___ Clear criteria in evaluating options?	_____
___ Alternative best met criteria?	_____
___ Were risks assessed?	_____
5. Gain Support and Approval	
___ All key players identified?	_____
___ Was their support assured?	_____
___ Were risks assessed?	_____
6. Develop Plan	
___ Includes clear action steps?	_____
___ Includes completion dates?	_____
___ Includes lines of responsibility?	_____
___ Milestones clearly defined?	_____
___ Risks identified and managed?	_____

Preparation has the major objective of ensuring quality control and accountability in the physical preparation of the Implementation and installation equipment – all of which relate directly to the work of step 2. The activities in this phase include the handling of all required actions, ordering, prototyping, testing of flow changes, and training: If handled according to specifications, this phase provides an audit trail and the necessary supporting documentation to ensure readiness for the actual implementation to begin.

Once key players and their concerns are identified, a roadblock analysis helps remove or manage those concerns prior to the final presentation and implementation.

Solution/Idea Description: _____

Key Player	Role				What will each team player like about the idea or solution?	What concerns will each team player have with the idea or solution?
	A	I	W	C		

A= Approves **I = Implements** **W - Works Within** **C = Champions**

Getting people to take the correct actions requires persistence and an ability to remove obstacles. During Implementation, follow-up with employees is important to demonstrate commitment. Revisit the methodology within Mobilization. Communicate what is about to occur to the entire organization. As the process changes and creates success, the rest of the organization will naturally get on board. Leaders communicate what is about to occur to the entire organization. As the rest of the organization sees successful process changes implemented, they will naturally want to join in the fun. Teams will check the pilot to:

- See how it meets the vision goals
- Determine the improvement in value stream gaps and process cycle efficiency and cycle time
- Ensure human behavior changes as the new workflow process is implemented

Step 3: Rollout and Continuously Improve

The improved process is now ready for full, uninhibited application and integration into the value stream. In step 3 – Rollout and Continuously Improve, teams fully apply the improvements to the process, document results, and identify the next steps for additional improvements. The team will:

- Follow the daily corrective action plans for implementation of critical actions.
- Manage Implementation so as not to overwhelm the people.
- Measure the improvement and visibly display the results of cycle time and cost savings.
- Aggressively pursue process qualifications to meet ISO standards and continually improve the process.
- Update the Center of Excellence master project schedules.
- Create a storyboard consisting of before and after pictures.

The goal is not to standardize the new process over the long term. The goal for Rollout and Continuous Improvement phase is to institutionalize a Continuous Improvement Culture environment. Leaders should seek ways to continually reduce cycle times and improve customer satisfaction. People will become improvement-focused if they are rewarded for it, and more important, if they are given the support and the power to do it. This is why the performance plans must be very clear. Once this happens, the culture is truly transformed into a learning organization.

A danger of the methodical approach described so far is that teams must be patient if assistance is needed from the Core Team for implementation. The Core Team may be helping several teams at once, therefore clear communication is critical. Teams should focus on the "Can Do" projects first. The Core Team can assist the Steering Committee in making decisions regarding priorities and can communicate this to the organization.

Organizations that hire outside help in the journey should view them as coaches, not ivory tower consultants. Coaches should not be afraid to roll up their sleeves and get their hands dirty. Too often programs fail because too much time is spent "planning" rather than actually demonstrating how to run the new workflow process. Many consultants "Talk the Talk," but few have what it takes to "Walk the Talk."

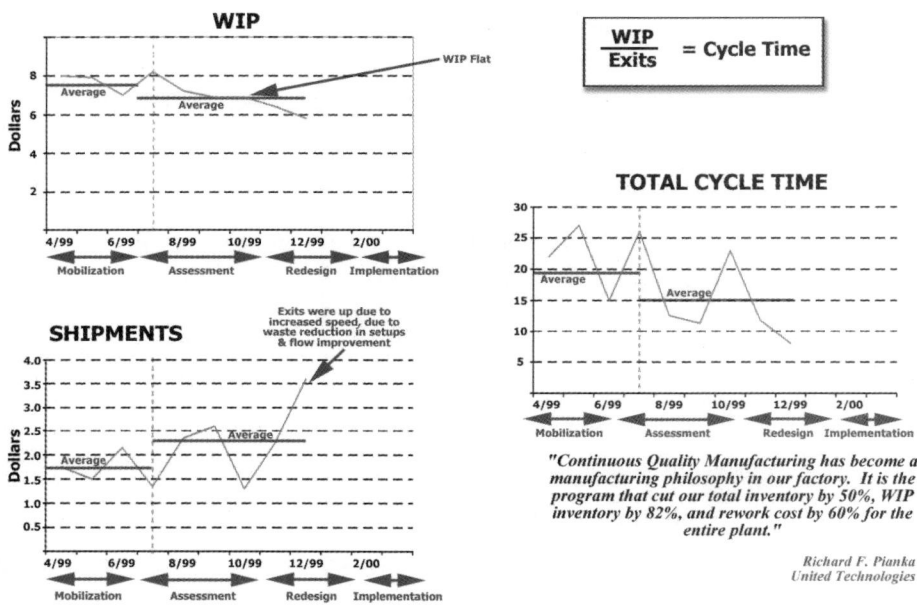

$$\frac{\text{WIP}}{\text{Exits}} = \text{Cycle Time}$$

"Continuous Quality Manufacturing has become a manufacturing philosophy in our factory. It is the program that cut our total inventory by 50%, WIP inventory by 82%, and rework cost by 60% for the entire plant."

Richard F. Pianka
United Technologies

At the end of the Implementation phase, the Steering Committee and the Core Team should have a review session to ensure that everyone understands the new workflow process. This session should be split into two parts. The first is the new workflow process techniques applied to reduce the cycle time of the process. The second describes the change management cultural techniques applied to help team members accept the new way of operating. Employees will have continuous improvement skills, knowledge, and the ability to repeat the methodology by this time.

To measure improvement outcomes:

- Analyze the measurements of new processes for sustainability.
- Create new accounting processes.
- Reward success by acknowledging all members.
- Communicate changes and raise the bar.
- Transfer the value of knowledge to the rest of the organization as lessons are learned.
- Communicate the success of the change.
- Repeat the process to further reduce non-value-added parts of the workflow.

Boundaries between other implementation programs and continuous improvement programs are often fuzzy and unpredictable. However, the boundaries between FlowCycle Implementation and continuous improvement are clear-cut, predictable, and repeatable.

The cornerstones of FlowCycle are to recognize, congratulate, promote, and reward people for a job well done. Highlight specific accomplishments so everyone throughout the company can see the Cycle Time Reduction results in the Center of Excellence.

This process must be in place before you set up your Implementation teams. Without recognition, people lose motivation, enthusiasm, and commitment to solve problems and make improvements.

Organizations can move quickly through MARI again and again by improving upon the very changes they just implemented. When executed according to the FlowCycle methodology, workflow process improvements gain momentum and support creating a straight path to World Class.

Implementation Summary

- Develop and implement employee training for new skills and knowledge (pilot and rollout).
- Implement new HR policies, procedures, and practices.
- Develop communication and conduct periodic Implementation.
- New goals for Cycle Time Reduction.

Deliverables:

- New performance plans based on Cycle Time Reduction.
- New measures throughout company.
- Follow corrective action plans for Implementation.
- Don't implement too much at one time.
- Measure improvement and visibly display.
- Continually improve process.
- Repeat the FMA Improvement MARI model.
- Advertise your success; organize and update your project notebooks.
- Update your CQM/BQM centers and display success in the Center of Excellence.

CONCLUSION

FlowCyle is a methodology for the twenty-first century. Many companies struggle today because they operate under twentieth-century assumptions. High costs are passed on to customers. If the product is slow to market, the customer simply has to wait. The response to a lack of improvement in a workflow process is more overtime for the employees. What do twenty-first century customers want? Robert Rodin answers that question in his book, *Perfect Free and Now*. Today's customer wants things perfect, now, and even though it's not possible, free.

Companies playing by the old rules are inflexible and unresponsive to internal and external environments and they organize themselves according to functional silos. This leads to a lack of innovative thinking and a lack of customer focus. Some of these companies have implemented Lean technologies with unsatisfactory results. A survey at the 2000 CEO Roundtable in Washington, D.C., found that only 39 percent of management teams were satisfied with their Lean efforts. A CFO quoted in an early 2000 *USA Today* article said, "Speed by fruitless cycle time reduction has become the number-one improvement tool to measure bottom line results to the shareholder value."

FlowCycle, the MARI Execution Methodology, is a system that goes beyond Lean and creates sustainable results through cycles of learning. It involves executive commitment and follow-through with strong project management. FlowCycle also uncovers and excises waste in non-manufacturing processes along with waste on the production line. Companies often find six times the savings in removing waste from business processes compared to manufacturing processes.

FlowCycle To Success

After observing the success of a FlowCycle implementation, Don Washkewicz, COO and President of Parker Hannifin Corporation, created its Corporate Awards of Lean Excellence. The awards recognize Parker plants, business processes, and organizations for enterprise-wide Lean improvements. The focus is excellence in Lean implementation, praising innovation, and highlighting the importance of strategic partnerships in the new age of technology.

Washkewicz thanked FlowCycle coaches for leading a journey that reduced 70,000 square feet out of a 300,000 square-foot plant, and reduced cycle time from 14 to two days. Parker's business-to-business Lean Enterprise allows company divisions to build Lean-to-Lean marketplaces, manage corporate vision, directly and indirectly source implementation materials, and connect users and service providers on the Internet.

The most rewarding aspect of a FlowCycle journey is that people are transformed. An aerospace facility implementing FlowCycle experienced resistance from a team member not wanting to attend meetings. The individual was a young man into rap music who had memorized over 140 songs word-for-word. One step of Mobilization is to create involvement by asking participants to memorize the three key principles, four phases and 13 steps of FlowCycle. An individual who can memorize over 100 songs has an enormous capacity for memorization and knowledge.

The young man turned the memorization exercise into a rhythmic recital and bought into the program. He became a key leader on a team that took a manufacturing line on which product traveled over six miles in a building of less than 200,000 square feet to a process traveling less than 200 feet. He became a supervisor in the process. FlowCycle turned this young man from a roadblock into a Cycle Time Reduction fanatic. He now applies MARI to improve both his professional and personal life. No one knows what is going to make people change. People react for different reasons and we should not prejudge people. Follow the process and the cream will rise to the top.

FlowCyle is built on the principle of sustainable results. Closed minds create stagnant processes that perpetuate waste. Managers that stop learning should stop managing. Managers that stop changing should stop leading. It is easy to get caught up in corporate politics and stop developing new core competencies. New core competencies create continuous improvement and these companies are leaders in today's international marketplace. The future of any company lies in the hands of its employees.

A FINAL WORD

Ultimately, who is responsible for change? You are. As an organization and an individual, you must look into the mirror and tell the truth about what you see. Do you see something that needs to be changed? Once you realize you have the ability to change your mind, you can become proactive about the future.

You are a reflection of your leadership, and your organization is also a reflection of your leadership. Employees need a sense of vision, purpose, and values. They need to be told they are important, how much they are relied upon, that they are trustworthy. They need to be given training, such as the FlowCycle methodology, and role models. Employees need to be given the resources that allow them to fight waste in both the organization and in their personal lives. It is important to realize problems are found in bad processes. People are the solution.

Remember, those that can't change their minds can't change anything in their life or their company.

APPENDIX

View of a FlowCycle Company

- Non-value-added activities have been eliminated from workflow processes.
- Value streams contain fewer process workflows.
- Customer requirements control all processes.
- Continuous FlowCycle improvement efforts prevent non-value-added activities from entering workflows.
- Workflow cycle times are shorter.
- Total quality permeates the business.
- The main flow of work is visible and understood by everyone, especially to those doing the work.
- The organizational structures are more linear and shallow.
- Boundaries within and between workflows have been simplified to a point that work passing between workflow processes encounters no barriers.
- Obstructions become obvious to everyone.
- Teams are self-controlling with little need for outside help.
- Faster workflow process cycle times necessitate rapid decision-making at all levels. This is accomplished by providing employees at all levels with the tools and skills required to make correct decisions.

- Leadership has no need to control day-to-day operations. Leaders are free to provide solid business planning and leadership by example and to grow the business.
- Strategies focus on collapsing the time-to-build value through time-to-customer-response and time-to-market for new products and services.
- The relationship between cycle time and workflow accuracy is exploded by exceptional competitive advantage. Companies with shorter cycle time enjoy rapid response to changing customer needs.

Those products and services are being built or serviced to customer orders, but within the customers' expected lead time. Responsively executed cycle times allow shortened forecast and planning horizons and will possibly alter the market.

- Quick gains are achieved by reducing the non-value-added time and simplifying the workflow process. This turns additional time into additional capacity and the ability to service the needs of a larger customer base. Every process in a company is part of a workflow to better service the end customer. A company's rapid cycle time execution enables the close alignment of the supplier into the company's processes.

GLOSSARY

Absorption Costing – An approach to inventory valuation in which variable costs and a portion of fixed costs are assigned to each unit of production. The measure usually is counterproductive to Flow/Lean implementations.

Accountability – Paired up based on personality styles. Accountability cannot be delegated, but it can be shared by people and by other team members. Developed by Dr. Tony Alessandra. Free behavioral style test at www.platinumrule.com.

Best Practices – Development of optimal standard operating procedures by gaining input from multiple process owners.

Business Quality Processes (BQP) – principle that reduces business process cycle time by the planned, continual elimination of inefficiencies and waste in business workflows.

Benchmarking – The third step of Assessment, which measures the company's products, services, costs, and practices against those of competitors or firms that display the "best in class" achievements to break paradigms.

Benchmark Measures – A set of measurements (or metrics) that is used to establish data collection for improvements in processes, functions, products, etc.

Continuous Flow Manufacturing – Engineering-based manufacturing solution used to identify bottlenecks and non-value-added expenses in order to increase productivity and lower inventory.

Continuous Quality Manufacturing (CQM) – On-going examination/improvement effort, which ultimately requires integration of all elements of the value stream of manufacturing processes to lower costs, defect occurrences, and delivery and cycle times.

Culture – An organization's external environment, influenced by such factors as geography, ethnicity, religion, creed, and education.

Cycle of Learning – Results after each MARI cycle completion, thus a continuous evolution and improvement of an organization's procedures. A process in which each of the individuals in the group is engaged in problem solving by using the MARI methodologies over and over again

Continuous Improvement Process – The fourth step of Implementation, a never-ending effort to eliminate waste from top down to root causes of problems.

Core team – A cross-functional team of specialists formed to manage FlowCycle introduction and processes to achieve the journey of Lean toward a Waste-Less Enterprise.

Corporate Accountability – The set of important behaviors that members of the company share. It is a system of shared values of what is important and of how the company works. These common assumptions influence the ways the company operates.

Critical Success Factors – Four groups on the Quad Chart sufficient for organizational objectives whose achievement should be sufficient for Lean success.

Cycle Stock – One of the two main conceptual components of Kanban/Pull Systems. The cycle stock is the most active component, i.e., that which depletes gradually as customer orders are received and then is replenished. The other conceptual component of the item inventory is the safety stock, which is a protection against downtime.

Cycle Time, In Manufacturing – the time between completion of two discreet units of production, e.g; the cycle time of converting raw material through final assembly of finished goods. In business workflow, it refers to the length of time from triggered need until the need is filled.

Cycle Time – Based on Performance Plans and Reviews – A corporate strategy that emphasizes time as the vehicle for achieving competitive advancement. Cycle time must be like a manufacturing cycle time plans and reviews of cycle time are measures of the length of the time from a triggered need until the need is fulfilled.

Commitment – In the FlowCycle philosophy, participation of all employees of the organization's improvement efforts. Participation includes accountability and serving on teams, deploying goals to reduce cycle time providing the resources and training that all levels need to achieve the goals, participating in quality improvement teams, reviewing organization-wide progress, recognizing those who have performed well.

Defect Prevention – Mistake-proofing techniques, designed in a way to avoid errors, which result in product defects.

Demand Flow – Processing raw materials with adherence to the guidelines dictated by customer demand.

Execution Methodology – see MARI.

Enterprise Value Chain – The sequence of customers, internal and external, form a chain of consumers.

FlowCycle Measures – A set of measurements (or metrics) that seek to establish the current level of performance of a workflow of a product, sector or department. Six baseline measures are usually established before mobilization activities.

FlowCycle Production Process – A workflow management philosophy that includes a consistent set of principles, procedures, and techniques in which every step of a value stream is evaluated in terms of value-added and non-value added.

Flowchart – A chart that shows the value stream operations or product families. FlowCycle maps are visuals to better understand workflow processes. The flowchart is one of the tools of Assessment.

FlowCycle Mapping Symbols – An analysis tool that graphically presents five simple symbols used to represent operations, data, transportations, inspections, storages and delay.

FlowCycle – Multi-dimensional change effort that goes beyond simply "Lean" by focusing every aspect of the enterprise on achieving speed in meeting customer requirements through an organized, systematic methodology. FlowCycle is a sustained, permanent day-by-day method, rather than a quick fix employed simply to meet desired financial numbers.

Gantt Chart – Horizontal bar chart graphically displaying progress in relation to time.

Just-in-Time (JIT) – A philosophy of manufacturing, based on planned elimination of all waste. It encompasses the successful execution of all manufacturing activities required to produce a final product. The primary elements of Just-in-Time were developed by Toyota in Japan.

Information System – Overall compilation of computer hardware and software, human resources, and processes used for planning, assessment, and management.

Information Technology – Collection of computers, communication systems, and any other device that provides the means for organizing, processing, and implementation of data, personnel, and methodologies.

Kaizen – The Japanese term for continuous process improvement.

Kanban Trigger – The Japanese word kanban means card. The term is often used synonymously for the specific scheduling system developed and used by the Toyota Corporation in Japan. In a FlowCycle environment it specifies what to build, when to build and how much to build.

Knowledge Capital – The accumulated knowledge for achieving a faster cycle of learning.

Lean Manufacturing – reduction of cycle time through elimination of non-value added waste.

Lean Accounting – A cost accounting system that measures costs based on value streams activities performed and then uses cost drivers to allocate these costs to products, customers, markets, or projects.

Lean Enterprise – A group of individuals, functions, and companies that operate in synchronized organizations. The workflow value stream defines the enterprise. It includes values, work ethics, education, and consumer and ecological factors.

Kanban/Scheduling – The process where the kanban triggers the schedule and adjustments are made to the schedule based on cycle time of production to ensure that inventory levels and customer service goals are maintained and using Takt rate capacity and proper material planning.

MARI – acronym representing FlowCycle's™ Execution Methodology: Mobilization, Assessment, Redesign, and Implementation.

Muda – Japanese word meaning waste.

Needs Assessment – The initial element of FlowCycle methodology, performed to determine the current state of affairs and to set goals for the future state.

Net Requirements – The on-hand inventory balance minus allocations and buffers held for quality problems, often called net available balance to reduce floor space.

One-Piece Flow – Concept of utilizing a seamless business/manufacturing process, such as an assembly line, thus eliminating non-value-added activities.

Pareto chart – A graphical tool to rank causes, from most significant to least significant. It is based on Pareto's law, which was first defined with respect to quality by J.M Juan in 1950.

Point-of-use delivery – Direct delivery of material to a specified location on a plant floor near the operation where it is to be pull from a kanban trigger.

Pull (system) – The production of items only as demanded or to replace those taken for customer requirements.

Quality Circle – Regular small group discussion of employees conversing over quality issues and concerns.

Reengineering – comprehensive overhaul of a corporation's processes in order to modify work functions which result in superior performance.

Supplier Value Chain – The functions inside and outside a company that provide services to the customer. The workflow of involving suppliers early in the product design activity and drawing on their expertise, insights, and knowledge to generate better parts.

Six-Sigma – Goal of achieving a success rate in which defects are only experienced 3.4 times out of 1,000,000. (99.99966% good.) The term is usually associated with Motorola, which named one of its key operational initiatives Six-Sigma Quality.

Set-Up Reduction – Process that aims to eliminate or diminish the amount of cycle time needed to set up equipment by shrinking the necessary resources.

5S's *(Sort, Set in order, Shine, Standardize, Sustain)* – philosophy based on five Japanese words beginning with the letter 'S'. Objective is to improve quality and safety by reducing waste and non-value activity, as well as simplifying work environment.

Takt Rate – Achieved by dividing the number of orders by the amount of available production time.

Total Employee Involvement – An empowerment side FlowCycle in which employees are to participate in actions and decisions making that were traditionally reserved for management.

Total Productive Maintenance (TPM) – Preventive maintenance plus continued efforts to adapt, modify, and refine equipment to increase flexibility, reduce material handling, and promote continuous flows. It is operator-oriented but involves all qualified employees in maintenance activities.

Total Quality Culture (TQC) – Foundation for creating an atmosphere of change by developing the knowledge of cost of quality within an organization and using that knowledge to drive change.

Total Quality Management (TQM) - Technique that draws participation of all employees to improve the organization's overall methodologies, products and corporate culture. A term coined to describe Japanese-style management approaches to quality improvements. The methods for implementing this approach are found in teachings of W. Edwards Deming.

Toyota Production System – System that allows stopping a production line to correct defects, as opposed to employing line inspectors and repair operators to make adjustments while keeping the line running.

Throughput – Amount of work that can be performed by a system or process in a given period of time.

Value Chain Management – The planning, organizing, and controlling of supply value chain activities. The workflow processes within a company that add value to the goods or services of an organization.

Value-driven Enterprise – An organization that is designed and managed to add utility from the viewpoint of the customer, in the transformation of raw materials into finished goods or services.

Value Stream – The processes of creating, producing, and delivering goods or services to the market. The value stream encompasses the supplier, the manufacture and assembly process, and the distribution network from order taken to account closed.

Value-Added/Non-Value-Added – Determined by whether a certain activity directly contributes to meeting customer specifications.

Value Stream Mapping – Visualization of the specific actions required in providing a product or service to customers. (Internal and external).

Waste-Less Enterprise – Organization free of any non-value added processes. Workflow-Series of processes, activities and tasks completed to meet customer demands.

Workflow Cycle Efficiency – A measure (usually expressed as a percentage) of the actual output to the value-added time. Efficiency measures how well something is performing relative to existing value stream. The standard is 40 hours and 2.5 hours of value-added time – 2.5/40 multiplied by 100%, or 97.5%.

INDEX

A

Accountability partners, 108–110
Alessandra, Tony, 108 Free test. www.platinumrule.com
"analysis paralysis," 132
Assessment. *See also* FlowCycle; Implementation; Mobilization; Needs
    Assessment; Redesign
    analyzing business for leverage, 152–153
    analyzing customer requirements, 133–136
    assessing current performance, 137–148
    benchmarking, 148–151
    managing resistance during, 154–155
    overview, 70, 71–72, 129–132
    summary, 155

B

behavior styles, 108–110
benchmarking, 148–151
Black & Decker, 87
Business Quality Process (BQP) centers, 105, 138–148
Business Quality Processes (BQP), 16, 52–53, 61–64
business value-added activities, 142. *See also* non-value-added activities;
    value-added activities

C

Center of Excellence (COE), 59, 118–119, 158
CEOs Conference of 2000, 48–49
change. *See also* Mobilization
    coping with, 54–58
    creating, 88–91
    Human Change Curve, 93
    responsibility for, 194
    understanding, 91–94
charts. *See also* maps
    Gantt charts, 179, 180
    Pareto diagrams, 116, 153
cherry-picking of jobs, 168
Chevrolet, 24
communication, 77
Compaq Computers, 2, 4
Continuous Flow Manufacturing (CFM), xii, 42, 46, 65
Continuous Improvement, 29–30

Continuous Quality Manufacturing (CQM), 16, 53, 65–67
Continuous Quality Manufacturing (CQM) centers, 105, 138–148
Core Team, 102–103, 135, 166, 173. *See also* Mobilization
cost of quality, 59–60
cost-reduction edicts, 129–130
Culture Assessment, 75. *See also* Needs Assessment
Culture Survey, 75–83. *See also* Needs Assessment
customer value-added activities, 142. *See also* non-value-added activities; value-
    added activities
customers
    external, 134, 135–136
    internal, 134, 136
Cycle Time Reduction, 5–7, 21, 27, 28, 46–48, 65–67
cycles of learning, 50–51

D
Defect Prevention, 12
Dell, Michael, 2, 4, 6
Dell Computer, 3–5
Demand Flow, 18
Deming, W. Edwards, 90, 164
Deming quality system, 15
"Dove" behavior style, 108–110
Dover Elevator, 154–155
Downtime, 39

E
"Eagle" behavior style, 108–110
empowerment, 79
Execution Methodology, 2
external customers, 134, 135–136

F
feedback and recognition, 81
5-S Program, 12
Flow Management & Associates, Inc., xii–xiii
FlowCycle. *See also* Assessment; Implementation; Mobilization; Needs
    Assessment; Redesign
    advantages, 44–50
    Business Quality Processes (BQP), 52–53, 61–64
    Continuous Quality Manufacturing (CQM), 53, 65–67
    cycles of learning, 50–51
    description, 7–9, 29

Lean Manufacturing versus, 11–19
MARI Execution Methodology, 26–28, 46, 50–51, 70
One-Piece Flow and variety, 25–26
origins, 41–42, 46
overview, xi–xiii
FlowCycle symbol, 9
Total Quality Culture (TQC), 16, 52, 53–60
FlowCycle Triangle, 70
Ford, Henry, 23–24, 25, 39, 41, 85
future-state maps, 170–172. *See also* maps

G
Gantt charts, 179, 180. *See also* charts
Get a Grip! (Capozzi), 15
Greenspan, Alan, xi–xii

H
Holtz, Lou, 121–123
Human Change Curve, 93

I
IBM, xii, 2–4, 17, 41–42, 46, 65
Implementation. *See also* Assessment; FlowCycle; Mobilization; Needs
        Assessment; Redesign
    developing detailed implementation plans, 179–181
    overview, 70, 72–73, 177–178
    pilot and rollout testing, 181–185
    rollout and continuous improvement, 186–190
    summary, 190
Implementation Teams, 105–110. *See also* Mobilization
internal customers, 134, 136
inventory, 37

J
JIT/Lean Manufacturing, 14
Journey Plan, 86–87. *See also* Needs Assessment
Just-In-Time (JIT), xii, 6, 12, 13, 23, 41–42, 48, 65

K
Kaizen Events, 12, 13, 14–15, 18, 42–43, 48, 65
Kanban/Pull Systems, 66–67, 84, 115–116, 158, 160, 162–163, 165–168

L

leadership
    Culture Survey and, 80, 82
    importance, 57–58, 95–96, 101, 124–128
Lean Manufacturing
    description, 10–11, 18, 25
    FlowCycle versus, 11–19
    shortcomings, 48–49
Lean Thinking (Womack), 11

M

maps. *See also* charts
    future-state maps, 170–172
    process flow maps, 115, 132, 141–142
    Value Stream Mapping, 12, 115, 132, 134, 137–148, 160, 166–168
MARI Execution Methodology, 26–28, 46, 50–51, 70. *See also* Assessment;
        Implementation; Mobilization; Needs Assessment; Redesign
Mission Task Statements, 99–101, 112. *See also* Mobilization
Mobilization. *See also* Assessment; FlowCycle; Implementation; Needs
        Assessment; Redesign
    conclusion, 121–125
    core strategies, 102
    Core Team, 102–103
    core values, 101
    creating change, 88–91
    Implementation Teams, 105–110
    launching change management efforts, 117–121
    Mission Task Statements, 99–101, 112
    off-site team building experience, 126–127
    organizing for success, 111–112
    overview, 70, 71
    Sector Teams, 103–104
    selecting processes for redesign, 112–116
    Steering Committee, 94–99, 106–108, 111, 120
    summary, 127–128
    understanding change, 91–94
Motorola, 12

N

Needs Assessment. *See also* Assessment; FlowCycle; Implementation;
        Mobilization; Redesign
    Culture Assessment, 75
    Culture Survey, 75–83

Journey Plan, 86–87
Operations Assessment, 83–85
overview, 73–74
net requirements, 165
non-value-added activities, 31–33, 45, 142–145, 153. *See also* value-added
    activities

O
off-site team building experience, 126–127. *See also* Mobilization
Olson, Brett, 87
One-Piece Flow, 23–26
Operations Assessment, 83–85. *See also* Needs Assessment
order fulfillment, 113–116
"Owl" behavior style, 108–110

P
Pagonis, Gus, 3
Pareto diagrams, 116, 153. *See also* charts
Parker Hannifin Corporation, 192
"Peacock" behavior style, 108–110
Perfect Free and Now (Rodin), 191
performance
    assessing current performance, 137–148
    performance plans, 172–175
PlyGem Industries, 47–48, 50
Powell, Colin, 98
process flow maps, 115, 132, 141–142. *See also* maps

Q
quality, cost of, 59–60

R
Red Flag Process, 120–121
Redesign. *See also* Assessment; FlowCycle; Implementation; Mobilization; Needs
    Assessment
    creating further redesign, 163–168
    documenting pilot for change, 169–175
    overview, 70, 72, 156–158
    setting design goals and priorities, 158–163
    summary, 176
Reengineering, 6, 12, 13, 18, 48, 62, 173
resistance, managing, 154–155
rollout. *See* Implementation

S

scrap, 40
Sector Teams, 103–104. *See also* Mobilization
Setup Reduction, 12
Shingo, Mr., 3, 6, 33, 41
Shingo #9 Engine Plant, 150
Shula, Don, 73
Site Champion, 103, 110, 111
Six Sigma, 12, 158
Steering Committee, 94–99, 106–108, 111, 120. *See also* Mobilization

T

Takt boards, 140
Takt rates, 23, 38, 84, 116, 166–168
Takt time, 140
Taylor, Frederick, 173
Team Training, 12
teamwork, 78
Total Quality Culture (TQC), 16, 52, 53–60
Total Quality Management (TQM), 12, 13, 18
Total Quality Management-III (TQM-III), 48
Toyota, 22–23, 25–26, 150
Toyota Production System (TPS), 12, 13, 23, 25, 41, 65
TPS/Lean Manufacturing, 14–18
transition. *See* change

V

Value Stream Mapping, 12, 115, 132, 134, 137–148, 160, 166–168.
    See also maps
value-added activities, 45, 142–145, 153.
    See also non-value-added activities
value-added time, 30–34
Volkswagen, 17

W

Waste-less Enterprise, 10, 18, 52, 73
Welch, Jack, 29–30
Womack, Jim, 11
Woods, Tiger, 45–46, 149
work in process (WIP), 36–37
workflow
    description, 19–22
    improving, 22–23, 47

One-Piece Flow, 23–26
    value stream and, 21–22
Workflow Cycle Efficiency (WCE), 30–40, 43
workflow management, xii–xiii